iTunes Music

Mastering High Resolution Audio Delivery

Produce Great Sounding Music with
Mastered for iTunes

Bob Katz

f Focal Press
Taylor & Francis Group

NEW YORK AND LONDON

First published 2013 by Focal Press
70 Blanchard Road, Suite 402, Burlington, MA 01803

Simultaneously published in the UK
by Focal Press
2 Park Square, Milton Park, Abingdon, Oxon OX14 4RN

Focal Press is an imprint of the Taylor & Francis Group, an informa business.

Library of Congress Cataloging-in-Publication Data
Katz, Robert A.
    iTunes music : mastering high resolution audio delivery : produce great
  sounding music with Mastered for iTunes / Bob Katz.
        p. cm.
    Includes index.
    ISBN 978-0-415-65685-6 (pbk.)
  1. iTunes. 2. Mastering (Sound recordings) I. Title.
    ML74.4.I49K37 2013

Typeset in ITC Korinna by Mary Kent
Printed in Canada

This book is dedicated to the genius of Steve Jobs,
who conceived the master plan.

Thanks

Thanks to my wife and partner, Mary Kent, who has contributed to this book in countless ways, creatively, supportively, spiritually, semantically, stylistically, graphically, photographically…

Bob Ludwig, of Gateway Mastering, Portland, Maine, has mastered more top-selling recordings than anyone can count, and he's the nicest guy in audio. He has made generous suggestions, checked and rechecked the iTunes facts, and helped make this book a lot more complete than it ever could have been. He was kind enough to provide us with the story of how **Mastered for iTunes** came about, which you will find in the Foreword that follows.

Thanks to my editor and good friend, Renaissance Man Chris Morgan, who I've known for 40+ years. Together we've produced magazine articles, record albums, CDs and finally got to do a great book. It's been fun. Let's keep on keeping on!

Thanks to Jim Johnston, who has generously checked the technical facts, contributed diagrams, test signals and all round good humor! No pun was left unturned, and no fact left unconsidered. But please blame me if you find any error—I probably forgot to run it by JJ! Send any notes of errors to our page at **www.digido.com/iTunes**.

Thanks to my spirited colleagues in the **Music Loudness Alliance**, Florian Camerer, Eelco Grimm, Kevin Gross, Bob Ludwig, and Thomas Lund, who work together to inform the music world on loudness issues.

Thanks to B.J. Buchalter of Metric Halo Labs for devising **Spectrafoo**, my favorite high-resolution audio analysis program. Thanks to Christian Mildh, who critiqued and helped take our cover to the highest kick!

My thanks go to Anaïs Wheeler, Focal Press Editor, who offered many helpful suggestions for the book's design.

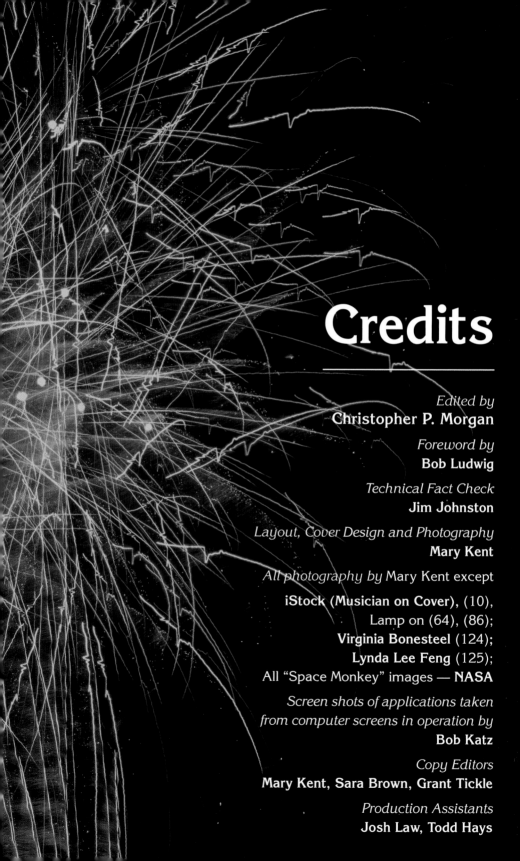

Credits

Edited by
Christopher P. Morgan

Foreword by
Bob Ludwig

Technical Fact Check
Jim Johnston

Layout, Cover Design and Photography
Mary Kent

All photography by Mary Kent except

iStock (Musician on Cover), (10),
Lamp on (64), (86);
Virginia Bonesteel (124);
Lynda Lee Feng (125);
All "Space Monkey" images — **NASA**

*Screen shots of applications taken
from computer screens in operation by*
Bob Katz

Copy Editors
Mary Kent, Sara Brown, Grant Tickle

Production Assistants
Josh Law, Todd Hays

Contents

Mastered for iTunes
The Untold Story

Today, there are hundreds of **Mastered for iTunes** (MFiT) titles and, for online music files, MFiT will be the standard from now on. It is 100% based on solid science as you will learn in the chapters of this book, but in this Foreword, you'll hear the "untold" story of how Mastered for iTunes came to be.

It evolved through the concerted efforts of many critical listeners both inside and outside Apple, Inc.

In 2004, I began to notice some music tracks that were difficult to encode into 128 kilobits per second (kbps) AAC streams. Seeking a solution, I contacted Bill Stewart, Engineering Manager with Apple's audio group. I told him about a tricky song that sounded poorly when I ripped it from my CD into iTunes.

He explained how AAC works: it defines a set of tools that you can use to **encode**. These tools must be implemented by the **decoder**. The **decoder** has a strict definition of how it deals with the data stream and how it renders the data generated by the various tools defined in the standard. The specification, however, does not describe exactly how the **encoder** is to create the bits it generates. Some of the tools are optional. Thus, although **decoder** operation is completely defined, the **encoder** still has some leeway to make qualitative choices. The goal is to achieve transparency between the original signal and the encoded signal at 256 kbps.

Let's fast forward to 2009, when I re-ripped from the same CD and listened to the results. I found that the quality of the AAC encode was noticeably better. It had crossed the line from really unacceptable to acceptable. That's when I realized that, quietly and behind the scenes, Apple had been incrementally improving the sound quality of their AAC encodes since its introduction with the opening of the iTunes store in 2003.

In 2008, top music producer and mixer Bob Clearmountain and Betty Bennett, co-founder and CEO of Apogee Electronics Corp., started the ball rolling with talks with Bill Stewart about the audio quality of downloadable music. "We were so impressed Bill would take our concerns seriously," Ms. Bennett said. "We had a passionate conversation about music and the importance of high quality archiving for future generations. Soon after that, Bill was on his way to LA with a team of incredibly bright mathematicians and engineers, intent on making the AAC audio format the best it could possibly be. They were so motivated, they spent three days doing critical listening and evaluations in state of the art studios where the music was mixed, mastered or recorded..."

Bill Stewart's team, including Frank Baumgarte, Senior Audio Scientist with Apple's audio group, conducted listening tests at Wonderworld Studio, Marcussen Mastering, Lurssen Mastering, Apogee's Studio, Mix This, Zeitgeist Studio and the Fox Newman Scoring Stage with a host of professional listeners.

Apple thus began an 18 month quest to investigate how to create the most transparent AAC encodes possible, and to improve the sound of all those existing player/software systems without the customer paying any extra. This included the introduction of 24-bit source encodes that the 32-bit float AAC encoder can now legitimately use as well as "variable bit rate" encoding which allows AAC to use more bits on music that it evaluates as needing extra bits. 256 kbps is thus a baseline target bit rate but it can go up 10-15% larger if the encoder thinks it needs to do that.

The next chapter in our story finds one of Apple's Senior Directors, Robert Kondrk, speaking with producer Rick Rubin about improving the sound quality of Apple's existing sonic landscape—with over 250,000,000 players out there, all playing 256 kbps 44.1 kHz music purchased from the iTunes store.

At about the same time, Robert met Maureen Droney through her friend, Music and Audio Consultant Lisa Roy. Maureen is the Senior Executive Director of The Producers and Engineers (P&E) Wing of the Recording Academy (the Grammy® organization). She offered the resources of the P&E Wing to assist Apple in any way possible to further the acceptance of this initiative.

Things came to fruition in June 2011 when famed Fleetwood Mac *Rumours* engineer/producer Ken Caillat was finishing up his daughter Colbie Caillat's new *All of You* album. He had an approved CD, but was unhappy with the AAC encodes he had created by ripping a CD reference. He wanted to sell high quality vinyl and high resolution digital files so he conferred with Maureen about the present state of higher quality digital delivery.

Through Jeff Greenberg at The Village Recorders, Maureen set up a meeting there with Robert Kondrk, Ken Caillat, Jeff Greenberg, and Lisa Roy. As the discussion started to get technical, Eric Boulanger, the engineer at The Mastering Lab in Ojai, CA who was working on the Colbie Caillat project, joined the group.

Eric realized that Robert Kondrk was on a quest to make iTunes more professional and more accepted by professionals and artists in the industry. Apple wanted to "right the wrong" of the 16-bit sources that were being used, when often much higher resolution files existed. So, for the first time, Apple allowed an outside engineer access to their in-house encoding tools, which would allow Eric to do a high resolution pass of the album.

Eric worked with Bill Stewart and his engineering team to further refine the encode process, abolish the unnecessary file conversions in the chain, and remove the "scale reduction" software that had been used. Scale reduction was an algorithm used to slightly lower the encode level to prevent encoder clipping. With its removal, Apple created *afclip* software which tells the mastering engineer exactly what clips the encoder is creating and how to precisely deal with their removal—a big improvement. Now those tools, and more, are available to everyone.

At last the moment arrived: Eric remastered the album in 24-bit creating an iTunes master that was approved by him, Colbie, and the producers, with the same sound-quality standards that are usually applied to the CD. These new references received extremely positive reviews from all the creators. The new master was uploaded to the Apple servers "by hand," outside of the mainstream ingesting system. Avery Lipman, from the Universal Music Group, helped guide the project through the new paradigm.

Colbie Caillat's *All of You* thus became the world's first **Mastered For iTunes** title to be released on the iTunes store on July 12, 2011.

Foreword

Shortly after, I was working on Coldplay's *Mylo Xyloto* album, and I too wanted the encodes to be as good as possible. Eric Boulanger, calling from Maureen Droney's office, told me about the initiative which led to my beta testing the final *afclip* software tool Bill Stewart's team had created. The Coldplay album was the first one to use the *afclip* tool to measure clipping as an artifact of the AAC encode. Coldplay, their engineers and management, were very excited about this.

Eric continued with all the Pink Floyd Catalog, and later spoke to Andrew Scheps who, with producer Rick Rubin, issued the Red Hot Chili Peppers album, *I'm With You,* which became the second MFiT title released on August 22nd.

Things were really moving along when the Producers and Engineers Wing began a big push to network the word about the initiative.

Many of Eric Boulanger's Pink Floyd **Mastered for iTunes** titles were released on September 23rd. Coldplay became the 4th artist with a release on October 21, 2011, the same day that I chaired a Platinum Mastering Panel at the Audio Engineering Society convention in New York about Mastered for iTunes to a packed house of engineers waiting to hear the message. And the message is that the system's improvement can range from perceptibly better than before to dramatically better.

One of the newest Apple tools allows for A-B-X testing. I guarantee that, in the future, many people will fail the test when trying to tell the difference between the high resolution source and the Mastered for iTunes encode.

Bob Ludwig, July 2012, Portland, ME

Introduction

iTunes Music is written for musicians, recording, mixing and mastering engineers, producers, program directors, audio enthusiasts, students, and computer enthusiasts. In short, anyone who wants to get the most out of mastering music for the iTunes medium. Whatever your role in the music-producing process, we'll give you the inside tips and tricks you need when preparing your submissions for iTunes.

The thrust of this book is **high resolution audio**, which complements Apple Computer's exciting new **Mastered for iTunes** (MFiT) initiative, introduced in early 2012. This is a major step forward in the world of audio: for the first time, record labels and program producers are encouraged to deliver audio materials to iTunes in a *high resolution format*, which can produce better-sounding masters. Working with high resolution audio requires a little more care and technical knowledge in production. After reading this book, you may not be a technical expert, but you'll be knowledgeable enough to explain yourself properly in a multi-vendor environment, and converse with co-workers, engineers, producers, label executives and employees, and distributors. You'll also be able to set some facts straight when you encounter formidable resistance to good sound.

iTunes Music introduces techniques that all quality-conscious producers of music should follow. Many of the practices will be familiar to you if you work in broadcasting, but the music world at large is still awakening to the important concepts of lossy coding, high-resolution audio, loudness standards, loudness normalization, and dynamic range control.

The chapters here cover both theory and practice. To get the best sounding recordings, all the members of your team, musicians, producers, label executives, mastering engineers, or any other interested parties need to work together. So be sure all the actors in your troupe are familiar with the "bits" covered here. Buy this book for your friends who have a need to know: they'll be glad you did!

What is iTunes, and How Did it Get Started?

iTunes is the world's most popular music platform. It debuted on Apple MacIntosh computers in early 2001, before the first iPod was introduced in November that year. iTunes was based on a music-only program published by software producers Casady and Greene, who sold it to Apple. Its developer, Bill Kincaid, went on to Apple to lead the iTunes development team.

Over the years, the development team expanded iTunes to support multiple formats (books, movies, pdf documents, podcasts, television shows, and videos) and integrated all these media categories with the iTunes store. By the third quarter of 2011, the iTunes store had sold over 16 billion songs, making it the largest online music store in the world. We tend to take this powerful application for granted, and too often forget that iTunes is **free** and runs on all Apple products and Windows operating systems. Hundreds of developers maintain iTunes and create new features for the consumer's pleasure.

After a plethora of piracy nearly killed the music industry, Apple's overture demonstrated that downloaded music could be cool, desirable and profitably sold. Sexy ads with earbud-wearing dancers injected a new vitality, the hardware sold the music, and vice versa. The iTunes store was the enabler, giving the music industry a new life. All musicians yearn to get their music on iTunes, even before they think of Columbia, Universal, Sony, EMI, BMG, or hundreds of other record labels that have come and gone.

Apple's Sound Quality

Apple has always been concerned about quality as well as ease of use. Apple's iDevices (iPod, iPhone, iPad) are not only easier to use: they have more headroom and less distortion than most other music players, so they can take advantage of better sound formats as they become available. As an Apple fan and owner of countless Apple computers and devices since 1984, I'm really proud to see the computer, the hardware, and the OS evolve to become a remarkable music machine.

What is High Resolution Audio?

High Resolution Audio, in a nutshell, is audio created with longer wordlengths than 16 bits (e.g. 24 bits). A higher sample rate than the CD standard 44.1 kHz is also desirable. When handled properly, high resolution audio can sound purer, quieter (less noise), more dynamic, more open, and more natural than standard CD audio, which we abbreviate here as "1644." Working in high resolution, a mastering engineer armed with good sources can produce the biggest, widest, purest, most impressive-sounding recording! Chapter 2, *The Resolution Revolution*, goes into this topic in detail.

Audio Mastering

While audio mastering is a featured topic here, the purpose of this book is not to teach you how to master music. You will learn about all the new things that a mastering engineer needs to know in this iTunes world, and some of the practices that a mastering engineer needs to follow to get great sound.

Mastered for iTunes and the Independent Artist

Mastered for iTunes (MFiT) is important to Apple, and they are promoting it heavily. It will be initially available only to major labels, but because of the massive interest and public demand, distribution services and aggregators like Tunecore and CD Baby will shortly be accepting high resolution files for conversion to iTunes Plus format. So be sure you obtain and archive the high resolution files from your mastering engineer. Submit the 1644 files, and ask your distributor when you will be able to resubmit the higher resolution files. Professional mastering engineers use the techniques described in this book to ensure that all master files are the best quality possible. As a result, professionally-mastered 1644 files contain much of the information from the high resolution source. But when these files are converted to iTunes format, they lose a little more than high-resolution sources, so make sure your high resolution wishes are known at your distributor! It won't be long before their action follows your demand.

Another approach is to distribute high resolution audio directly via retailers like HD Tracks, who may be interested in selling your 2496 files at their online audiophile store. Releasing encoded (AAC) files on iTunes and simultaneously releasing lossless high resolution material on HD Tracks for audiophiles and sound purists produces a powerful parallel effort.

AAC Format

iTunes can play various audio formats, including WAV, AIFF, mp3, ALAC (Apple Lossless) and the focus of this book, **AAC (Advanced Audio Codec)**, the lossy-coded format used by all files that are sold on the iTunes store. As you will learn in Chapter 3, *Lost and Found*, and as Apple knows, AAC files produced properly from high resolution sources do sound better, which is the genesis of the MFiT program.

Links and Forum

Further information about any company, service or internet source mentioned in this book can be found at Digital Domain's support section for this book: **www.digido.com/iTunes**. This not only prevents "link rot," but also provides an entry page to our new forums, where you can ask questions and interact with other program producers working to make the best out of the iTunes format. You can also find updates, news, and book errata at this page.

Mini-Glossary and Main Glossary

To help you get through the new terminology, every chapter has a mini-glossary at the end, covering just the terminology introduced in that chapter. There is also a full glossary at the back of the book. In the E-edition, the mini-glossary is replaced with an electronic glossary in the standard format for your E-book platform (most E-books permit clicking on the word to find its definition). The comprehensive alphabetical index at the back covers every major topic within the chapters.

Summary of Chapters

Chapter 1: *Assembling The Album—iTunes Style.* This chapter looks at the contrasts between iTunes production style and Compact Disc production. The CD was introduced before the Personal Computer revolution. The PC enabled the transformation to iTunes, our approach to media, and our orientation from physical product to download. What is metadata? How is it used? Where is it stored? How do we get metadata to the distributor?

Chapter 2: *The Resolution Revolution.* What is high resolution audio? How to maintain resolution. How to destroy resolution (if you're not careful)! The sonic effects of resolution loss. Dithering, and why we dither.

Chapter 3: *Lost and Found.* An introduction to Perceptual Coding. Drawing parallels between audio production and photography.

Chapter 4: *Keeping It On The Level.* What is distortion? What causes distortion? Is there such a thing as "good distortion?" When does "good distortion" become bad? What is headroom? Crest factor? Loudness Range? What is True Peak level?

Chapter 5: *Loudness Normalization.* Why loudness normalization will eventually come to music production, as it has to broadcast. iTunes *Sound Check*, its performance today and tomorrow.

Chapter 6: *How Loud is Loud?* Loudness metering. How the new loudness meters make our music work easier.

Chapter 7: *Tools of the Trade.* The primary audio processing tools used in mastering and how we can use them to produce better and more consistent-sounding product. Taking advantage of Apple's Mastered for iTunes tools.

Welcome to *iTunes Music!*

www.digido.com/iTunes

Even Ajax can tell when it's been Mastered for iTunes!

CHAPTER 1

Assembling the Album iTunes Style

The combination of the iPod, iPhone, iPad and the iTunes software has changed both the way record labels release music and the way people buy and listen to it. iTunes' greatest asset, the playlist, lets listeners shuffle music or combine songs from different albums and artists at will. iTunes has also changed the status of the traditional record album, which for decades has dominated the music scene. Today, although most artists still release songs in the album format, the single is King.

iTunes vs. CD

When creating masters for both CD and iTunes, be aware of these basic differences:

1) **Singles:** Every iTunes song becomes available as a single as well as part of an album. This is a bonus, because making iTunes singles is far less expensive than manufacturing CDs: it costs a lot of money to produce, press, release, and market a physical CD with a single song on it. Before iTunes, record labels and producers would think twice about releasing a CD single. Today, making a single is a common practice.

2) **Master format:** The master for a CD album consists of one audio file, in a proprietary image format, which includes all the gaps between songs, and segues (overlaps) between songs. In contrast, the master for an iTunes album consists of a series of WAV or AIFF files, with the pauses (gaps) between the tunes included at the end of each audio file.

3) **Metadata:** The metadata for a CD is built into the CD master's subcode (the subcode is hidden data that controls the player's display and operation). Metadata includes: ISRC codes; track marks; pause marks; EAN barcode; CD text (album and track titles, genre, artist); and potentially much more, including graphics, though CDs with graphics are very rare. Mastering engineers produce and error-check the CD masters, and encode this metadata. When there is no time for the producer/A&R to personally approve the final reference, the mastering engineer will typically be asked to proofread any changed metadata and then send the master to the duplication plant. The master format can be a simple physical CD, which the plant replicates, automatically transferring the metadata, or a DDP fileset sent to the plant via ftp, containing all metadata.

By comparison, the individual WAV files delivered to iTunes do not contain metadata. Do not try to encode metadata into these files as it may prove confusing and useless as iTunes will not use it. So the artist or producer must deliver metadata separately; they can no longer depend on the mastering engineer to encode metadata. Metadata for an iTunes album must be entered at the online digital distributors' websites, or, for larger record labels, into forms that can be submitted direct to iTunes. The person responsible for ensuring that this data is accurate, correctly spelling the titles and artists' names, and submitting this data, has a tremendous responsibility. If you are an artist or producer submitting information, have as many people as possible check the data before submission, use copy/paste to minimize possible typos, and proofread many times over. Changing information that has already been submitted to iTunes is a time-consuming and frustrating process. Proofread that data! Then proofread it one more time. "Measure twice and cut once."

Not all distribution services accept all metadata. Some may not accept composer or movement information for classical releases.

4) **Pause marks:** Pause marks on CDs indicate the gap between tunes. These gaps can also contain audio intentionally excluded in shuffle play. Audio in the gap will not make it to the radio either (unless the radio station plays past the pause mark). It is common to use the CD gap for spoken introductions, count-offs, breaths, applause, laughter, and ambience (in live albums). The CD gap also can be used as a hidden track, heard only by listeners playing the entire album.

But there are no **pause marks** in iTunes, and they may be gone forever. In iTunes, all the "in-between" content appears at the tail of the

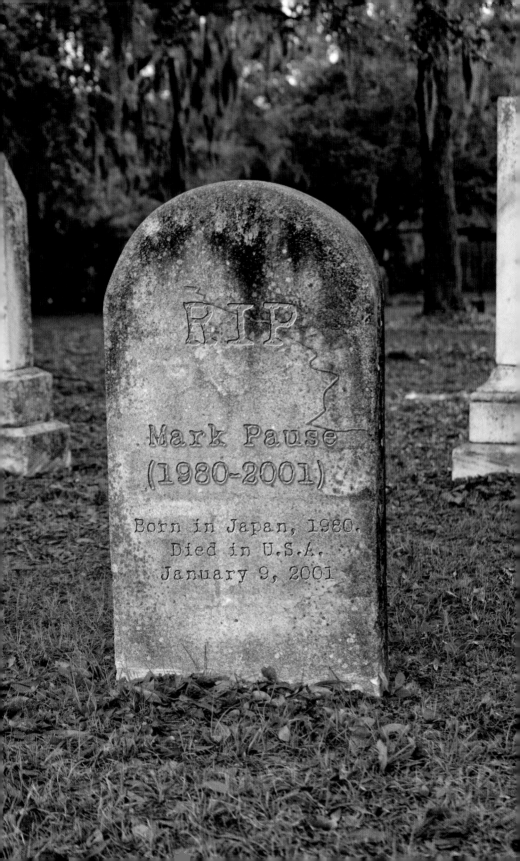

previous tune, and it will be heard in shuffle play (that can slow down a party). This makes artists hesitant and a bit less adventurous, so you're less likely to find creative or interesting extras between the tracks of an iTunes album.

A related change brought on by iTunes is that the song length includes the gap length, so royalties charged by the licensing agencies (e.g. Harry Fox) increase, since they license songs by their length. This can get costly with live albums. If you're low on cash, you could create a song or two called "applause." But then iTunes will charge anyone who downloads it—scratch that idea!

5) **Segues and Gapless Playback**: Gapless playback on CD is simple and built right into the format. It permits segues (overlaps or crossfades) between tunes. When the CD was King, listeners were not confused cuing to a song and hearing the fade of the previous song mixed over the head of the next. iTunes users can rip singles from their CDs, then put them into their own playlist; but CDs with segues make it hard to extract just one song for multi-album playlists.

iTunes, version 7 and later, and the more recent models of iPods both support gapless playback, as long as *crossfade playback* is turned off in the preferences. If *crossfade playback* is turned on, then the checkbox called *part of a gapless album* must be checked for each individual tune. A bit confusing, but most people do not turn on *crossfade playback*.

Now that every song can be downloaded as a single, artists have become increasingly reluctant to create an album with segues. They fear that listeners will not understand a song that does not "complete" (i.e., come to a well-defined stop or fade out), or whose beginning contains part of the end of the previous track. In the image on page 17, the track mark indicates both the beginning of the next track and the end of the previous, which is where the first song would be cut off when it is downloaded as a single. A listener playing from this track mark will hear both the end of the previous song and the beginning of the next. When I master an album and suggest a segue to the producer, he often rejects it for these reasons. But crossfades display colors in the artist's creative palette—the artist's right to be different! A good solution is to create additional singles that do not contain the segues, also known as the "Radio Version." Since albums have a fixed price, let's include the single versions as bonus tracks—though when you hear excerpts from *Dark Side of the Moon* on the radio, stations will usually play the album version and fade out during the segue. Now that broadcasters can use music from iTunes, will they discover the Radio Version bonus tracks?

Your Tracks Are Now in the iCloud

Apple, Inc. announced their iCloud service on June 6, 2011. If an iTunes music customer signs up for an iCloud account, then purchases a track on iTunes from one device (e.g. iPhone, iPad), it will automatically download on up to 10 devices or computers that use the same iTunes store account (for example, another Apple computer, iPhone, iPad, iPod Touch, etc).

CD Text and Gracenote

When you insert a CD in a modern car player, it displays the artist and track titles. This feature, called **CD Text**, is part of the metadata encoded by the mastering engineer. iTunes can write CD Text, which becomes useful when cutting a custom CD for the car. But iTunes is not able to directly read CD Text from an audio CD. The only way to transfer metadata from a CD is to use a third-party applescript (which can be found at our links page) to import the information into iTunes' database.

Apple decided not to use CD text for a very good reason: a lot more data is available about artists, music and genres than can fit on an audio CD, and there is no way to keep this data current—what if the band changes its name? iTunes gets its metadata through an online database run by Gracenote. If your computer is online when you insert a CD, within moments the CD is identified and you see the artist, title and other information. When iTunes is not connected to the Internet, it may read this information from its cache if the CD was previously inserted in your computer.

Beware of Identical Timing

Gracenote identifies the CD by the song lengths, pauses, and distribution of the song timings. But because many CDs have been produced in the past few decades, it's possible for two completely different CD albums to have the identical timing information. That's why it's important for you or the mastering engineer to confirm that there is no identically timed CD already in the Gracenote database before sending the master for replication. Insert the CD reference into iTunes and make sure it does not come up with someone else's name. To guarantee that iTunes performs the Gracenote inquiry, choose **Advanced/Get Track Names**. If it displays another CD title, ask the mastering engineer to minutely change the gap or length anywhere on the CD, which will usually fix the problem. If Gracenote has registered more than one album with identical timings it will usually display a list of candidates from which you must pick. Incidentally, every possible CD single has already been registered in Gracenote's database, so your new CD single will display the wrong title until you register it. This can cause a great deal of confusion!

How to Submit Metadata to Gracenote

There are two ways to submit metadata to Gracenote: the slow, unreliable way, and the dependable, secure way. The slow way is to insert the master CD into iTunes and choose **Advanced/Submit CD Track Names**. This could take weeks to months because Gracenote is overwhelmed with such requests from around the world. If you misspell any piece of metadata, there is no direct way for you to correct it except to try to submit the names again. Imagine how that must confuse Gracenote! And Gracenote will not contact you about any submissions. Another reason to avoid this method is that anyone can take your pressed CD and submit bogus artist or song information (I leave the possibilities to your imagination).

The most reliable method to submit CD metadata is through the **Gracenote Content Partner Program**. Major record labels and some mastering houses are members of this program; they will securely encode the metadata and submit it to Gracenote on a priority basis. This locks out any malicious users from trying to change your metadata. There is also a mechanism to revise and resubmit the data, should a spelling or other error turn up. This correction propagates quickly through the Internet.

A Big Day at the Mastering House: Mastering for iTunes, CD and LP

Most artists continue to press CDs in addition to offering their music as digital downloads. Sometimes mastering engineers have to make three masters—for LP, CD and iTunes—of the same album, each with its unique technical requirements. Here's my approach to dealing with multiple masters:

1) If the client's source files are not already at double sample rate, upsample all the client's mix files to 3296 using Weiss's Saracon

software. If the client's files are already at 88.2 kHz, we work at 88.2 kHz sample rate. (Other mastering engineers may effectively "upsample" by using DACs and ADCs as described in Chapter 2).

2) Place these files into a 96 kHz Playlist (*VIP* is the term that Sequoia uses. Other Digital Audio Workstations (DAWs) may use the terms *EDL, Session, or Project*). Track marks and CD Text information can be entered into Sequoia at any step in the process. I take advantage of that by getting all the metadata from the client in advance (if possible), placing track marks and entering the CD Text into the 96 kHz Playlist.

3) If any of these files are supposed to overlap another, I create the segue at this point, then listen and adjust until the client and I are happy with the sound, as pictured below.

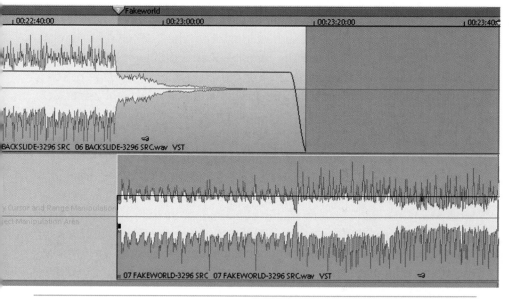

In this Sequoia Playlist, each track is stereo, but we can view only one channel to save space and make things easy to read. The top track contains the end of song #6, which overlaps the beginning of song #7. Note the track mark at the beginning of song #7. The album version of song #6 does not complete, and song #7 contains the end of song #6 as well as the beginning of #7.

4) Next, I apply any desired digital processing available in the DAW to the source elements. I then feed the processed audio signal from Sequoia via a digital route to any external digital or analog processors. To feed analog processors, I first convert the signal to analog, then, after processing, back to digital again. Working at double sample rate minimizes the losses of conversion, and the higher sample rate warms up the sound by reducing alias distortion when clipping or compressing.

I apply different compressor, equalizer, or other processor settings to each song so it gets its own proper treatment and level. The return from the external chain then gets recorded into a new Sequoia track to a *capture* file.

5) When songs are segued together it is often valuable to capture each song without the crossfade for radio or single versions. During the mastering session, when I encounter a segue (as in the figure on the previous page), I mute the second track and capture the end of the first "clean" into its own WAV file. Then I go back before the beginning of the second track, mute the first, and capture from that point on. Now I've created two independent capture files that can later produce individual single WAVs or be combined to produce the segue. This technique also leaves an escape route if we change our minds about the segue.

6) The individual capture files next get converted to 3244 via Saracon and are brought in sync into a 44.1 kHz Sequoia Playlist, preserving the track marks and metadata from the original 96 k Playlist. Different dither shapes affect the tonality of a recording in different ways (dither is explained in Chapter 2). For the CD, I choose a shape of 16-bit dither that best fits the musical source, and cut either a physical CD master (to be shipped to the replication plant) or a DDP master that can be directly uploaded to the replication plant via FTP.

7) For iTunes, the mastering engineer exports one WAV file for each song. Conveniently, Sequoia can name the exported WAVs according to the name of the CD track. The end of each exported WAV file must incorporate the gap between tracks at its end, or include any segue up to the next track mark. If the client is in the Mastered for iTunes program, I examine the entire album for encode clips, and I may drop the level of the album until there is no more AAC clipping (more on how to avoid AAC clipping in Chapter 7, Tools of the Trade), then export each WAV to a 2444 file, which will be sent to Apple for encoding. Not all clients are aware of, or worry about, the sonic compromises of clipping and loud levels when encoding, but I try to enlighten them on this point. If the client is not in the Mastered for iTunes program, I explain why dropping the level to remove AAC clipping produces better-sounding AAC files; I then drop the level if they agree, and dither the signal to 16 bit while making individual 1644 WAV files. These 1644 files are sent to Apple, who perform the encoding. Another possibility is to send a physical CD master, but I prefer to have more control over the integrity of the master and directly produce the WAV files for distribution.

8) The mastering house is not responsible for typos in CD Text. To help the client with their proofing, I upload a special digital CD reference to the client to check it with a DDP Player program designed by Sonoris. They can examine the artist, title, ISRC, EAN and other metadata information and sign off on its accuracy. They can then cut their own CD ref to check the CD text in the car and evaluate the sound on various systems.

Every record label, streaming service, and distributor has its proprietary metadata database—it's a tower of Babel out there. Sometimes we receive a spreadsheet from the label with all the information they require. I once designed an auto-import macro for our database, but it was only suitable for one record label. We use copy, paste and four-eyeball-proofing around here. It would be ideal to be able to import or export metadata from Sequoia in a standardized format (e.g. XML), but there are too many competing XML formats. For now, clients have to enter data by hand for each distributor or streaming service they subscribe to, in addition to Gracenote.

Keep in mind that English is the standard language of CD Text masters. Officially, CD Text must be in the ISO-8859-1 character set, which includes accented characters. But in practice, accented characters can produce garbled text in a car CD player. So I am generally conservative and avoid accented characters. Disobey this advice at your own risk!

9) In a perfect world I would master the dynamics separately for each destination medium: Give the LP and CD a more liberal dynamic range, not subject them to the squashed dynamics of a song destined to be played on iPod ear buds at the gym. But it's not practical to make two different masters except possibly in the case of the LP, as I can run a 96 kHz stream during mastering which does not have any peak limiting. I always try to make as dynamic a product as possible, still taking into account the noise of all the destination media (including the car). In the future, I hope that iPods and iPhones incorporate smart compressors that measure the noise of the environment. Then mastering engineers could produce a single "audiophile" master, the one that sounds great in a quiet living room or home theater and it would also play well in the car. Wouldn't that be nice!

16-bit fixed   This is the standard fixed-point wordlength of the Compact Disc. 16 binary bits express a coded range of 96 dB, but with care and proper dithering, a 16-bit file can capture signals as low as -115 dB below full scale. The difference between floating point and fixed point notation is explained in more detail in Chapter 2, *The Resolution Revolution.*

1644   The abbreviation used throughout this book for 16-bit, 44.1 kHz audio.

2444   The abbreviation used throughout this book for 24-bit, 44.1 kHz audio.

3296   The abbreviation used throughout this book for 32-bit float, 96 kHz audio.

44.1 kHz   The sample rate of the Compact Disc. This means that 44,100 samples of audio are captured in each second.

96 kHz   A higher-resolution sample rate often used for high-fidelity music recordings. This means that 96,000 samples of audio are captured in each second.

A&R, A&R Director   Artist and Repertoire. A title often given to the production director at a record label. His or her job is to work with and develop the artists signed to the label.

AAC   Advanced Audio Codec. Also abbreviated as AAF (Advanced Audio File format). Files which have been encoded as AAC may have one of these extensions: .m4a, .mp4, .3gp (the latter is used in cell phones).

ADC   Analog to Digital Converter.

AIFF, aiff   Audio Interchange File Format. The file extension is .aif. This was the file format most commonly used at Apple throughout the 80s and mid-90's but it has been largely replaced by WAV because WAV files (specifically Broadcast WAV files) can contain metadata while there is no metadata standard for aiff. Other than the internal numeric format, there is no audible difference between AIFF and WAV and the two formats can be interconverted with absolutely no loss.

Alias   See Alias Distortion.

Alias Distortion   An unwanted form of beat note, the result of interaction between the sampling frequency and an original signal (including the distortion components of the original signal), resulting in frequencies that are not present in the original signal. Any signal over half the sample frequency will yield alias distortion below half the sample frequency, and this distortion is most often neither harmonic nor musically consonant. For more information, see Chapter 4, *Keeping it On the Level.*

Barcode   The barcode is a unique code that identifies product for sale. Normally, it identifies the entire album, but sales organizations have begun to use the album barcode as part of the product identification for selling singles derived from that album. But the album barcode is not enough: Each vendor or store must establish additional accounting codes besides the barcode to keep track of individual sales of singles from the album. The EAN barcode (International Article Number, formerly known as European Article Number) has 13 digits. The UPC (Universal Product Code, developed in the U.S.) has 12 digits. To maintain upward compatibility, older 12-digit U.S. UPC codes can automatically be converted to a legitimate 13-digit EAN code by adding an extra 0 at the left. However, all new 13-digit EAN codes used in the U.S. do not have to begin with 0! The checksum is the last digit to the right: it is a calculated digit used by barcode readers to protect against read errors. If you receive an 11-digit code from a distributor, it is a UPC without the checksum digit. 12-digit codes can be confusing. A 12-digit code could be a UPC number with a checksum, or the first 12 digits of an EAN without the checksum. Visit the digido links (see page 8) and navigate to the barcode check site to help clarify the situation. The site can calculate a checksum digit that you can compare with the last digit of the number you were given. When in doubt, check with the supplier of the barcode.

CD   Compact Disc.

CD Text   Metadata which is embedded in a CD containing artist, title, genre and other information.

Clipping   Digital Overload. The level of a digital signal cannot exceed full scale. When the gain would cause the signal to exceed digital full scale, the output becomes a squared-off wave, which looks "clipped," with moderate to severe distortion for the duration of the clipping. The output medium (e.g. DAC) can overload even if the sample peak of the file never reaches full scale. That's because many devices produce higher output levels than their input, including filters, DACs, and codecs. So the sample peak level of the digital file should be somewhat below full scale to yield an output or playback that does exceed full scale (clip).

DAC   Digital to Analog Converter

DAW   Digital Audio Workstation. Some DAWs popular with mastering engineers (because of their unique features) include: Pyramix, SADiE, Sequoia, Soundblade, and Wavelab.

DDP, DDP Fileset   Disc Description Protocol. An image file format developed by Doug Carson and Associates for use in manufacturing CDs, DVDs and certain other disc formats. It contains audio, track and metadata information, which the plant can use to cut a CD-Audio master for replication.

Dither   A special kind of noise which is used to reduce wordlength without adding distortion. See Chapter 2.

Double Sample Rate   Shorthand for audio files recorded at either 88.2 kHz or 96 kHz. Likewise, quadruple sample rate can be 176.4 kHz or 192 kHz.

Dynamic range   The difference between the softest and loudest average (not peak) signal, expressed in decibels (dB). Since a moment of silence before the music begins could be counted as contributing to a wide dynamic range, the official measurement unit of macrodynamic range is LRA (Loudness Range, see Chapter 5, *Loudness Normalization*).

EDL   Edit Decision List. The layout section of a DAW where audio files may be placed, edited, faded, or digitally processed. EDL is a term used by SADiE. Pro Tools calls this a Playlist or Session; Sequoia calls it a VIP (Virtual Project); the principle is the same.

FTP   File Transfer Protocol. A standardized method for transferring large files that avoids using a web browser with its limitations.

Gracenote   A private company which maintains a database of artist, song and artwork information accessible by applications such as iTunes or Windows Media Player.

ISRC Code   International Standard Recording Code. A unique identifier for each song or version of a song. The ISRC code is part of the metadata. Each recording of a song gets a unique registered code that is used to track playback on the radio and keep track of royalties. While U.S. law does not currently allow for performance royalties, other countries do use the ISRC code to pay the performers. An ISRC code is supposed to remain as an identifier of a song throughout its lifetime, even if its owner changes (e.g. if one record company acquires the assets of another).

Level   A measure of intensity, voltage or energy, expressed in volts, dB, power or other units.

Mastered for iTunes   (Abbreviated MFiT in this book). A program or initiative created by Apple, Inc. in February 2012 that permits for the first time high resolution files to be sent to iTunes for encoding, and provides a set of recommended guidelines. See Chapter 7, *Tools of the Trade*.

Metadata   Literally, "data about data." If we consider the sound of the audio or music to be its main data, then metadata is additional information about the audio file other than its audio. This extra data may be contained within the audio file's header (the so-called id3 tag) or in a separate database. Examples of metadata include album graphics, artist, title, composer, recording date, genre, and loudness information. In an audio CD, metadata is embedded in the subcode hidden within the disc's structure.

Pause mark   In the Compact Disc, an optional subcode mark which defines the pause between tracks. However, audio can be put in the CD during the pause yet remain identified as "the pause," making it useful for a few features which iTunes did not inherit from the CD, as described in this chapter.

Playlist   An iTunes document that contains a list of songs in the order the user wants them to be played.

Producer   The individual(s) assigned to help create the sound of an audio production and work with the artist(s).

S

Segue   Pronounced "seg way." From the Italian for "to follow," means to connect two songs together in an artistic way, usually by crossfading or blending the songs during the transition, sometimes by connecting the two songs instantaneously or abruptly.

Sequoia   A DAW produced by a software company called Magix.

Subcode   Hidden code within the Compact Disc that contains information in addition to the audio.

Track mark   A coded location in the CD subcode that indicates when a new track begins. There are no track marks in iTunes, the beginning of the audio file is the beginning of the track.

T

W

WAV   Pronounced "wave." The file extension is .wav. The most commonly used source file format in digital audio (AAC is probably the most common file format overall). It is a lossless linear pcm format and may contain metadata in its header, the metadata standard defined in the Broadcast WAV specification.

Zippy

UNLATCH COVER
WHEN NOT IN USE

CHAPTER 2

The Resolution Revolution

Whoever said that digital audio is much easier to deal with than analog never looked under the hood of a recording console! There are many steps involved in recording, mixing, mastering, and finally delivering the material to iTunes, and without some guidelines, it's all too easy to inadvertently lower the quality of digital music files. Fortunately, you can preserve the integrity of your digital music files by following the simple *workflow* steps outlined here. These "best practices" are recognized by recording studios around the world, and will help maintain the quality of your files even after significant processing as well as guarantee the best possible translation to iTunes AAC files. This is especially true now that Apple is accepting high resolution source files in the Mastered for iTunes program.

Wordlengths

The first consideration in preserving the quality of your digital music files is to maintain wordlength and data integrity of our standard PCM files. Each bit of the integer binary sample is worth nominally 6 dB (more accurately, 6.02 dB). In other words, a 16-bit sample can describe up to 96 dB of signal level difference (16 bits x 6 dB). A 24-bit sample can express a much greater range, up to 144 dB (24 bits x 6 dB). This form of binary is called *fixed-point binary,* also known as *integer binary.* In order to code even larger and smaller numbers, *floating-point notation* uses exponents and decimal points (technically speaking *radix points*) to be able to express very large and very small numbers with 32 bits. But since minus 96 dB is already quieter than the noise of the quietest room, why should we bother with even longer numbers?

Mathematical Range		
16-bit	**24-bit**	**32-bit float**
96 dB	144 dB	Greater than 1500 dB!

The answer is this dirty little secret of digital audio: Wordlengths expand. Any time we change gain, equalize, compress, or otherwise process a source, its wordlength expands. This is mathematically unavoidable. For example, if you start with a 16-bit source and change its gain as little as 0.1 dB (or any amount!), its wordlength will grow to the internal resolution of the DSP, which means, with most modern-day CPUs, that it will grow to a 32-bit number. Some dedicated DSP chips operate at even higher internal resolutions. Pro Tools HD will expand the wordlength to 48-bit. Weiss Gambit processors work internally at 40-bit. This means that the low level portion of the material has been expanded into the longer internal wordlength of the processor. Until further processing, the sound retains the same resolution and signal-to-noise ratio as the original file. But any processing, such as added reverb, equalization or compression, will be calculated and stored with higher precision in the longer wordlength. Using more accurate (i.e. longer wordlength) numbers means there will be less noise and distortion. The ambience, space, depth and purity of tone of your music have been spread into this longer wordlength, so important information would be lost if the lower bits were simply cut out (*truncated*). So the first key to good sound must be: *Don't cut out the lower bits,* or, *Bigger is better!* To avoid introducing *truncation distortion*, the file should ideally be stored at 48 bits (for Pro Tools HD), or 32-bit float (for DAWs such as Apple Logic, Nuendo, Samplitude, Sonar, or Pro Tools HDX, etc.). But there's no such thing as a 48-bit file, and many systems, including older Pro Tools systems, cannot use 32-bit float files. The solution, as explained below, is to first apply dither, *then* shorten the wordlength of your file.

Dither

Proper *dithering* means better sound for the end product, and critical listeners and clients can hear the difference. Dither is a special form of intentional noise added to files to eliminate the creation of distortion during file processing. Once processing, mixing or mastering is done, we usually have to shorten the wordlength, perhaps produce 24-bit fixed point files for systems that can't use 32-bit files, or 16-bit files for proofs or for the Compact Disc. The proper way to reduce wordlength is to first add the dither (in the correct amount for the shorter wordlength), then round the result to the shorter desired wordlength (the DAW performs this procedure "behind the scenes" when the user decides to dither).[1] Without dither, permanent *truncation distortion,* which is correlated with the signal, would result. Attempting to fix truncation distortion by adding noise later doesn't work—it's like closing the barn door after the horse has escaped.

Is there any disadvantage to dithering? A little bit of noise (which is unavoidable) is added, but the net result is a great reduction in distortion, so dithering is far preferable to simple truncation. Noise increase is about 3 dB, meaning 16-bit dithered audio's noise floor is 93 dB below full scale, and 24-bit is 141 dB below full scale. In an analog console, low level signals are gradually obscured by the noise floor, without added distortion. A dithered digital console performs much like its analog counterpart, revealing low level signals that are as much as 20 dB below the noise floor. With a signal analyzer, or the ear in a perfectly quiet location, we can detect sounds that are as low as 115 dB below peak signal level in properly-dithered 16-bit audio. So dithering increases the low level resolution and dynamic range by as much as 20 dB. Without dither, the signal peters out at about -85 dB, losing itself into a crackly-sounding, gritty distortion, so a lot of ambience is lost.

Dithering is sonically important even when reducing a 32-bit or 48-bit word to 24-bit. Some might say, *"How can that be, when 24-bit dither noise is 141 dB down from the top? No one can hear that."* While it's true that no one can hear dither noise at 141 dB below full scale, we *can* hear the results of *not* dithering. The distortion products that would result from simple truncation are much louder than dither noise, louder even than the noise of a typical room. These products may occur at frequencies far removed from the original frequency, and thus are not easily masked. Anecdotally, many engineers and producers can perceive an improvement in space and depth when using Pro Tools HD's dithered mixer, which works at 48-bit and dithers to 24-bit (compared to the non-dithered mixer, which truncates at 24-bit). Claiming a sonic improvement with dither noise at -141 dB may seem hard to believe, and the improvement may sound subtle to untrained ears, but it is quite audible in a good room with a good DAC. In summary, take care when mixing or mastering to ensure that all long word digital signal paths employ dither, even when reducing to 24 bits. In a complex chain where the signal passes from one process to the next, it only takes one undithered process or signal to create an unpleasant effect. Like chicken soup, 24-bit dither can't hurt and it will probably help! Pro Tools HD's dithered mixer automatically takes care of the dithering to 24-bit, but many other DAWs require that the dithering be set up in the preferences or introduced in plugins. Pro Tools HD (48-bit) has been superseded by Pro Tools HDX (32-bit float). The principles remain the same but the application and approach is slightly different.

As usual when it comes to audio topics, misinformation and controversy abound on the net about there being no need to dither the output of a floating point DAW's mix bus, although I am convinced that the improvement is audible. One argument presented for not dithering is that 32-bit float has only 24-bit resolution and that normal operations should not generate an out-of-bounds (32-bit) signal. The former is true, but the latter is not. The laws of physics are immutable: calculations (multiplications) create additional bit depth, and that requires additional bits. To demonstrate, I performed the following measurement which can easily be repeated: I generated a -120 dBFS 24-bit dithered 1 kHz tone, sent it through Sequoia (a 32-bit float DAW), and measured the result, shown below. The signal path to and from Sequoia is limited to 24 bits, so all the information in this measurement is strictly contained within a 24-bit signal.

First I passed the test signal through Sequoia operating at unity gain (shown in red). The result is clean, with only the 1 kHz tone visible plus the dither noise which was originally used to linearize the test tone

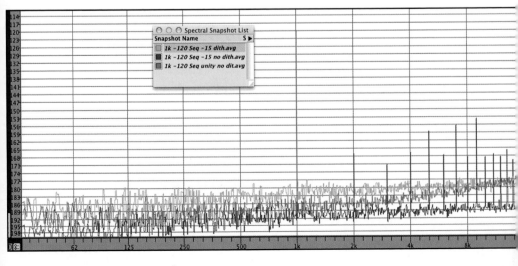

generator. This has a 24-bit noise floor, which measures -141 dBFS RMS across the whole spectrum.[2] Next, I dropped Sequoia's fader by 15 dB, which simulates a mix move (shown in blue), but without adding dither. As you can see, this generates distortion products throughout the high frequency spectrum. Then, by simply adding dither within Sequoia (shown in green), the distortion can be eliminated with the slight cost of about 3 dB more noise.[3] Where did this distortion come from? It obviously was caused by truncation of additional data (in bits 25-32), or dithering would not have been able to clear it up. In order for the Sequoia dither generator to work properly, it should calculate dither in 32-bit floating-point, taking into account the required fixed-point level of the dither.

"All Playback is Fixed Point (Integer)"

Although this distortion is measurable, the consequences of a single undithered attenuation may not be that audible. But don't forget that in a mix or even a complex mastering session, many faders, plugins, processors and complex routes are mixed together, each signal adding its own distortion, until the sum total becomes audible—unless dither is employed. Bear in mind that the distortion components from the undithered signals will then be distorted themselves, and at a relatively high level compared to the first set of distortion products; the results get exponentially worse if not dealt with. For an entertaining proof, drop a full scale sine wave at 997 Hz by .1 dB, look at the spectrum, then drop that by another .1dB, truncating to the original wordlength each time. Enjoy. Or not.

It is impossible to audition a 32-bit signal directly. All DAC inputs accept up to 24 bits, but they will truncate the bottom bits of an incoming 32-bit signal. In other words: All Playback is fixed point (integer). Because of this, if we try to compare a 32-bit float file with its 24-bit dithered result, the 24-bit will sound superior to the 32-bit! Truncation accidents can easily take place in the studio, because oftentimes the dithering settings are hidden in a menu.[4] But DAWs are not the only devices that can suffer from improper dithering of course: Digital monitor speakers, digital preamplifiers and receivers, monitor controllers, computer system and program volume controls all need to

---

Measurement of a 32-bit float DAW (Sequoia) using Spectrafoo FFT. 1 kHz -120 dBFS dithered test tone passed through Sequoia at unity gain, no added dither (shown in red). Drop Sequoia fader by 15 dB without adding dither (shown in blue). Add 24-bit dither in Sequoia (shown in green).

be dithered in order to preserve space and depth. We've come a long way, but we still have a way to go—for example, there are some monitor controllers on the market which are not dithered.

Storing Files for Later Processing

We can postpone the application of dither by storing intermediate products from DAW calculations as 32-bit float files, which do not require dithering until later reduced to fixed-point. Floating point lets us "have our cake and eat it, too." We can reopen a 32-bit float "intermediate" file, pick up in the middle of our work without adding noise or losing resolution, and send it to another DSP process. This is fine for *internal* DAW calculations or plugins, but the engineer needs to consider that *external* processors (either analog or digital) are limited to a 24-bit fixed point pipeline, so we have to dither when feeding into them. Furthermore, a well-designed external digital processor (such as a reverb unit, compressor or equalizer) should employ 24-bit dither on its output to the outside world. You should find this choice in its menu. Files which capture the output of external processors need only be 24-bit (no additional dither is needed). There is no harm in capturing the output of external processors into a 32-bit file, but it takes up unnecessary space.

Back to the Future

A properly dithered 16-bit CD (or file) made from a long wordlength source retains much of its space and depth. So what's wrong with having a 1644 master? Nothing at all, but it should be used only for producing Compact Discs. 16-bit masters make good final products, but should never be used as sources for further processing. Do not apply *any* changes to a 16-bit file: that means no gain change, no pan, no EQ, and no other processing. Instead, go back to the longer wordlength source. 16 bits is just at the edge of causing perceptible noise, and since any processing will add calculation noise (mathematics works), you don't want to do more work at the 16-bit level.

Avoid dithering too many times in a row. The sequence of dithering to 24 bits, doing another calculation, then dithering to 24 bits again, can be repeated several times in a row without audible degradation, but be careful about it. In comparison, dithering to 16 bits even twice in a row may reduce sound quality or audible resolution, because the dither noise may accumulate enough to obscure low level information. It's a subtle sonic loss, and you won't always hear it, but avoid cumulative dithering to 16 bits if possible. If you must use a 16-bit file as a source for remastering, there will probably be other sonic sacrifices besides cumulative dither. Chances are this file was already mastered, and contains the cumulative distortion of the original mastering chain. (This should ideally have been a very good chain, but nothing is perfect.) Compressors add distortion. One compressor may sound great, but two or three in a row may not. Double or triple cumulative distortion doesn't sound very good, unless you're trying for a special effect. Any compression will amplify the original dither noise by the amount of gain makeup in the compressor, plus the addition of a new layer of dither noise. Doing this at 16 bits can be a disaster. So there are many reasons not to use an already mastered file as a source—if you try to further manipulate it, the sound will go downhill fast. This is not a purist or audiophile distinction — everyone knows that returning to the source is preferable to working with a second generation. This is why compilation discs often sound worse than the original album: the 16-bit masters are often recycled, then processed. Remember: archive those sources. If you make a big hit, you'll need them again some day!

Higher Higher Higher

The argument in favor of using higher sample rates is less obvious than that for longer wordlengths. For example, the human ear cannot hear above 20 kHz (most people's hearing rolls off somewhere above 12 to 15 kHz), which means that 44.1 kHz is probably a more than adequate sample rate (the Nyquist theorem teaches us that cutoff frequency must be no higher than half of the sample rate). But there are distinct technical advantages to processing at double or higher sample rates, even if the destination sample rate is going to be 44.1 kHz. Clipping or distortion at higher rates produces less aliasing distortion, so digital processors running at double sample rates sound more "open." My

> **Here's a tip:**
>
> Include the wordlength and sample rate in the file name of every file you generate. E.G. "Love me do 1644 mast.wav." It's a lot easier than having to use a file checker every time you find a file. It will already be properly identified.

converters sound better to me at double rates, with diminishing returns above 96 kHz, so I generally work at 96 kHz. If you send 44.1 kHz or 48 kHz mixes for mastering, your mastering engineer may choose to upsample them first to take advantage of cleaner-sounding processing at the higher rates.

At the end of the process, mastering engineers working at the higher rates produce either 32-bit float or 24-bit fixed files at, say, 96 kHz, which is considered to be very high resolution. For CD, they would downsample to 3244 using the highest quality sample rate converters, then dither to 1644 using the best dither they think is appropriate for your project. But with the Mastered for iTunes Program, the engineer can send 24-bit files at single or double sample rate to the destination you specify (to you, your record company or your distributor). In the past, there has been an argument for working at exactly double rate (e.g. 88.2 kHz), but modern-day good SRCs (Sample Rate Converters) can convert between non-integer ratios without introducing noticeable problems. If you still believe in working at 88.2 kHz, there is no harm done.

Who Should Do The SRC?

Apple requires 24-bit files, so please dither your 32-bit float premaster files to 24-bit before delivery. Still, hold onto the 32-bit intermediate file, as it is the best origination format or source for further processing. For example, if it is necessary to perform any last minute level adjustments (or processing), using the 32-bit file avoids two stages of dithering. When Apple receives your 24-bit master file, they first archive a copy in Apple Lossless format using the Apple Lossless Audio Codec (ALAC), at the original sample rate and wordlength, to future-proof your submission (maybe someday iTunes will sell higher resolution ALAC files). If your file's sample rate is higher than 44.1 kHz, they downsample it using their mastering-quality Sample Rate Converter to a temporary intermediate 3244 file, then encode it to iTunes Plus format with their AAC encoder. However, if your masters are not at a higher rate and your mastering engineer did not work at a higher rate, do not upsample your source files. You will not gain any resolution and may possibly lose sound quality. Just send the master file at its original rate and wordlength.

Mastering engineers are by nature obsessed with producing the best sound. I've been relying on Weiss's superb **Saracon** SRC for all my heavy-duty rate conversion. But now that Apple has produced a master-quality competitor, it is hard to decide between using Saracon, sending 44.1 kHz files to iTunes, or letting Apple do the downsampling themselves. In other chapters we'll discuss other considerations, but in general either of these approaches should yield great results and everyone will benefit.

Our Workflow: From Mix to Finished Product(s)

Now that we know what's under the hood of our digital systems, let's put this knowledge to work to outline the steps from mixing through mastering and then preparing the master files for delivery. The diagram on page 35 describes one possible workflow where the end goal is to produce master files for LP, CD, iTunes (standard), Mastered for iTunes, iCloud, Internet download, and streaming. It's quite a comprehensive list! The file-naming conventions are my suggestions—use whatever approach you wish—but I do suggest that file names contain dates or version numbers, wordlength information, and indicate if the file is a master, mix, or other work part, because it's a tough job to keep files identified. I suggest using the word *master* only for final products that should not be further manipulated and are ready to be converted directly to the distribution format (e.g. AAC). Use the terms *source, mix, work part,* or *premaster* (for example) to name files that will be further manipulated before conversion to final product.

Let's follow the progress of a mythical song called "You Break My Heart" (not my problem).

Mixing (shown in red)

The original tracks (e.g., drums, trumpet, piano, vocal) are mixed and processed (pictured on page 35). The mix engineer may use one or more reverb chambers (an internal plugin or an external digital or analog unit), internal or external equalizers, compressors and other processes. External processors may be either analog or digital: if analog, then a separate D/A/D path is necessary. He may mix through an external analog console or summing mixer (not shown). The principles are always the same: whenever there is any processing (gain change, EQ, compression, etc.), wordlength grows, and wherever the wordlength must be reduced, dither is necessary on that output. Inputs do not need dithering, only outputs. Dithering inputs can deteriorate the sound. Notice how the aux send to the external reverb chamber must be dithered to 24 bits, and the output of the external reverb must also be dithered within the reverb. If dither is not in the external processor's menu and you cannot confirm that it's "built-in," then you are likely not getting the potential depth and dimension from the processor or chamber.

If a signal is summed digitally (mixed within the DAW), usually you should capture to a 32-bit float file. If you must capture to a 24-bit file, the output feeding that file should usually be dithered, except in the rare cases where the mix engine is internally dithered, as is the case of Pro Tools HD's dithered mixer.[5] If the signal is mixed via an external analog summing box, each output of the DAW should be dithered on its way to the summing box, unless the mix engineer never changes gains or adds any plugins on the way to the summer (a very rare occurrence these days). Refer to the file names at the right of the mix section for the possibilities of your particular DAW. In some cases it is optimum to mix to a 32-bit file, in others a 24-bit file is adequate. Again, there is no harm in mixing to a longer wordlength than necessary, but the reverse is not the case!

Mastering (pictured in blue)

Like the mixing engineer, the mastering engineer will probably go through the following steps (pictured on page 35):

Use a 32-bit float mastering DAW, along with internal and/or external processors.

Upsample the file obtained from the mix engineer, producing the 3296 file shown at the left, making it possible to proceed at a 96 kHz sample rate. Or (not shown), some mastering engineers play the original mix file through a superior-quality DAC directly to analog processing, followed by capture with an ADC at 96 kHz sample rate (which is a form of "analog sample rate conversion") **or** at 44.1 kHz, if there is no digital processing after the analog stage.

Level and/or process the sound, using plugins or internal processes, which work at 32-bit wordlength.

Feed an external processing chain, which has a 24-bit pipeline, so ensure the DAW outputs feeding the processing chain are dithered down to 24 bits.

Confirm that each external digital processor's output is dithered to 24 bits on its way to the next processor in the chain.

Optionally, capture the raw output from the external processors without limiting to a "prelimiter" file for use in cutting a more dynamic

Workflow: Mix, Master, Final Files

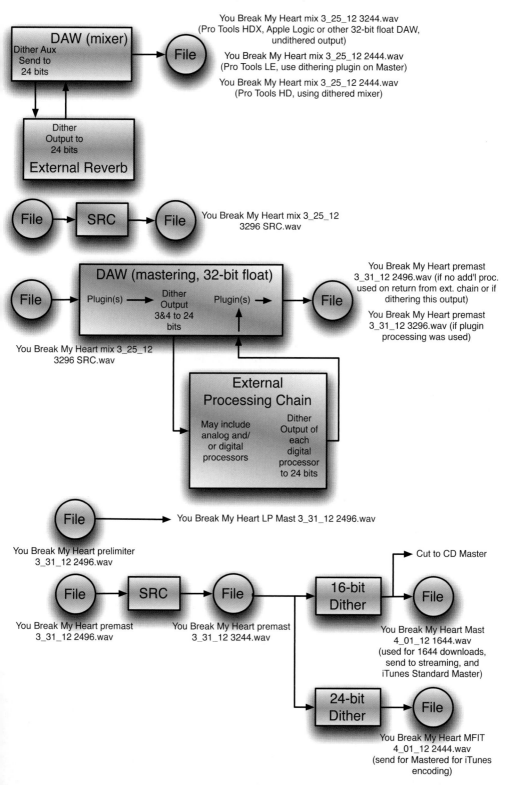

DAW (mixer)

Dither Aux Send to 24 bits

File

You Break My Heart mix 3_25_12 3244.wav
(Pro Tools HDX, Apple Logic or other 32-bit float DAW, undithered output)

You Break My Heart mix 3_25_12 2444.wav
(Pro Tools LE, use dithering plugin on Master)

You Break My Heart mix 3_25_12 2444.wav
(Pro Tools HD, using dithered mixer)

Dither Output to 24 bits

External Reverb

File → SRC → File

You Break My Heart mix 3_25_12 3296 SRC.wav

File

DAW (mastering, 32-bit float)

Plugin(s) → Dither Output 3&4 to 24 bits → Plugin(s) →

File

You Break My Heart premast 3_31_12 2496.wav (if no add'l proc. used on return from ext. chain or if dithering this output)

You Break My Heart premast 3_31_12 3296.wav (if plugin processing was used)

You Break My Heart mix 3_25_12 3296 SRC.wav

External Processing Chain

May include analog and/ or digital processors

Dither Output of each digital processor to 24 bits

File → You Break My Heart LP Mast 3_31_12 2496.wav

You Break My Heart prelimiter 3_31_12 2496.wav

File → SRC → File → 16-bit Dither → File

Cut to CD Master

You Break My Heart premast 3_31_12 2496.wav

You Break My Heart premast 3_31_12 3244.wav

You Break My Heart Mast 4_01_12 1644.wav
(used for 1644 downloads, send to streaming, and iTunes Standard Master)

24-bit Dither → File

You Break My Heart MFIT 4_01_12 2444.wav
(send for Mastered for iTunes encoding)

LP. The producer and mastering engineer decide which of the two files they prefer for LP, without limiting or the limited master which became the CD.

Optionally, insert a plugin for peak limiting on the return from the external processors.

Capture the output of the DAW to a "premaster" 96 kHz file (or whatever naming convention suits the workflow), with file name and filetype examples shown at the right side of the mastering section.

Final Files (pictured in green)

The prelimiter file (if chosen), is likely suitable for LP without further work (pictured on page 35). For clarity, the mastering engineer may rename this file as the LP master, as shown. The premaster file must be down sampled, first to a 3244 file. At this point the signal branches off depending on the destination requirement. For example, it needs to be dithered down to 16 bit for CD, for 1644 download, etc. As shown, it also needs to be dithered to 24 bits for Mastered for iTunes, where it will be sent to the record company to log and produce the information for metadata, and thence to Apple for encoding to AAC files and sale at the iTunes store. After this long journey, the music finally makes its way home to the consumer!

Now we're ready to discuss the concepts of lossy coding and AAC files, in the next chapter, called *Lost and Found*.

---

1 Jim Johnston wants to remind DAW designers to pay attention to truncating properly after dithering, by taking the largest integer smaller than the floating point number. Rounding after dithering is also safe. Improper or incorrect truncation would involve taking the largest absolute value smaller than the floating point number, with the right sign—which is very very bad. For us mortals, trust but verify: Use a DAW of high integrity designed by a company that you trust. Verify its integrity with measurement software.

2 The numeric scale at the left of the analyzer which goes down to -198 dB, is not a lie. It shows us that the actual value of each frequency component of the noise is much lower than the total noise floor. When all the frequency components are added together, they sum to a higher value, for example -141 dBFS. This often causes confusions when using FFT analyzers. Each band contains a small portion of the total noise; this does not mean that you have less noise, only that you have measured it in narrower bands.

3 This sensitive measurement tool reveals a few other interesting points: It shows the original dither noise of the test tone signal has been attenuated by the fader. The new dither

noise adds to the old. The shape of the dither used in Sequoia is flat, but the original test tone's dither has a rising high frequency characteristic.

4 In the book, *Mastering Audio*, I discuss some methods you can use to verify the data integrity and wordlength precision of your own DAW.

5 Pro Tools HD's dithered mixer is a special plugin that many mix engineers are not aware of, and it must be manually set up. Yes, it sounds better!

24-bit   The fixed-point wordlength that modern ADCs are capable of capturing, which means they can code a signal as low as 144 dB below full scale, but with noise-floor considerations in real rooms and the thermal limitations of components, it's likely the lowest real-world signal able to be captured by a converter is -120 dBFS (dB below full scale), though some converter designers are claiming as low as -130 dBFS.

32-bit float   This is the most common floating point wordlength used for calculation by CPUs and DSPs. The difference between floating point and fixed point notation is explained in more detail in this chapter,

3296   The abbreviation used throughout this book for 32-bit float, 96 kHz audio.

48 kHz   The most common sample rate used for digital video recordings, also used for original professional audio recordings. This means that 48,000 samples of audio are captured in each second.

Codec   Coder-decoder. In general, the purpose of a codec is to lower the bitrate of the source audio so it can fit in a smaller space and be transmitted faster. Sometimes referred to as "data compressor," this term should be avoided, as it can be confused with audio dynamics compression. Just call it a "coder" or "codec" and the audio file "coded audio." It may sound strange at first, but it's the only unambiguous way to communicate. A *lossy codec* reduces the amount of information from the original source (via psychoacoustic means). A *lossless codec* does not reduce the amount of original information, so there is a limit to how low the bitrate can be reduced in a lossless codec. Examples of lossless codecs include ALAC (Apple Lossless) and FLAC (Free Lossless Audio Codec).

Compression   Reduction of dynamic range, reducing variations in level, which can be performed manually (e.g fader ride) or with a processor (compressor). Compression of macrodynamics is often performed by moving a fader up and down, while compression of microdynamics is done using a processor to reduce the short-term crest factor.

To avoid confusion, please do not use the word *compressed* to apply to coded audio. Reserve compressed to apply to dynamic range reduction.

CPU   Central Processing Unit. The core processor in a computer.

D/A/D   Digital to Analog to Digital. Serial conversion from one format to another and back again. For example, in order to process a digital signal with an analog equalizer, the signal must be converted to analog through a DAC (Digital to Analog Converter), then returned to analog through an ADC (Analog to Digital Converter).

dBFS   Decibels below full scale. Often expressed with a space, e.g. dB FS. For example, -12 dBFS is 12 decibels below full scale digital.

Dither   A special kind of noise which is used to reduce wordlength without adding distortion.

DSP   Digital Signal Processor. Specially designed microprocessor that digitally affects the audio put through it. Depending on the algorithms designed for each specific chip some processes could include equalization, compression, limiting, bass management, etc.

Fixed-point A notation method of digital audio, with a fixed upper limit of 0 dBFS (full scale), and a fixed lower limit depending on the wordlength.

Flat No equalization. The "EQ curve" for flat audio is a straight line.

Floating-point A notation method for digital audio, with an extremely high upper and lower limit. For example, 32-bit floating point digital audio can express level differences of thousands of dB, which is only useful in the virtual world of calculation, not in the real world of audio which has at most 130 to 140 dB of dynamic range, and in practicality only 50 or 60 dB.

High resolution Audio which has a greater wordlength than 16-bits. A higher sample rate is also desirable. See **Resolution**.

PCM Pulse Code Modulation. The standard lossless method used to capture and reproduce digital audio.

Quantization distortion Distortion which can occur when converting between analog and digital or between two digital formats. This is not inevitable if dither is properly applied.

Radix point In simple terms, this is the same as "decimal point" for non-decimal notation. For example, in floating point, a dot, the radix point, is used to separate the exponent from the mantissa.

Resolution A rather loose term which in this book we use to define the wordlength and sample rate of a file. For purposes of definition, the lower the level of audio that we can "resolve," and/or the higher the highest audio frequency in the file, the better the resolution of the file. It's easy to say that a file is "high resolution," but a lot more difficult to prove it! If the file has been processed using high-quality techniques, then its resolution has likely not deteriorated significantly. The presence of high frequency information or of measurable low level information implies the file may be high resolution but not necessarily. Vice-versa, a recording with information "only" up to 15 kHz may be for many reasons a higher "resolution" recording than another which goes up higher. In the end, only the audio engineer knows the provenance of a complex chain of audio.

RMS Root-Mean-Square A means of calculating the average energy of a signal regardless of its waveform, in other words, despite the signal being time-varying, non-sinusoidal, or aperiodic.

Sample Rate Converter, SRC, also known as SFC (sample frequency converter) A device or software application that converts between one sample rate and another. Quantization distortion or loss of information can result if the SFC is not performed well.

W

Wordlength The number of bits required to define an audio sample or "word." For example, 16 bits are required to define each CD word, the wordlength of the Compact Disc is 16 bit. Same as "bit depth."

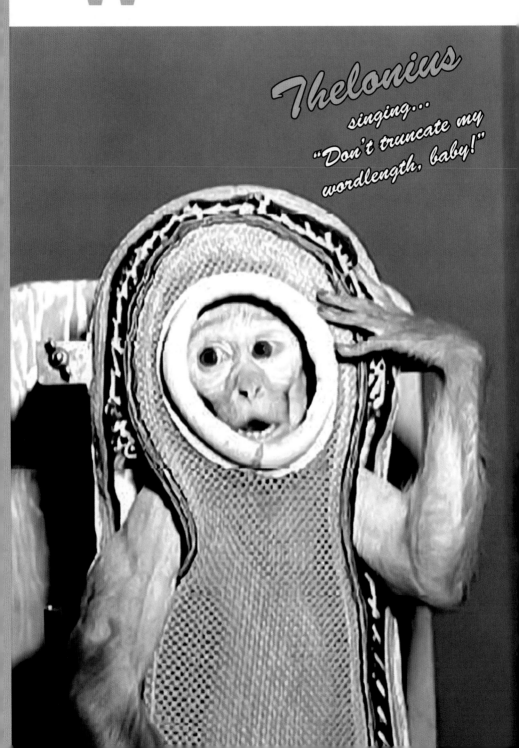

Thelonius singing... "Don't truncate my wordlength, baby!"

CHAPTER 3

Lost and Found

The Story of Lossy Coding

iTunes files are encoded in AAC format, which, like mp3, is an example of a *lossy coded* format. Lossy coding was originally developed to produce smaller-size files and decrease download time for an impatient consumer. Lossy coding is a complicated process: in order to make a smaller file size, a **Codec** (coder-decoder) discards some of the source data based on frequency masking, level masking, and other criteria. Some information is lost compared to the standard PCM we've been discussing up to now.

Sharp as a Photograph?

Digital cameras and photography are everywhere these days, and it's useful to compare the workflow of the professional digital photographer and the professional audio engineer. *The professional photographer* shoots *RAW format* photos. These original format photos losslessly capture the native resolution of the camera's sensor. But RAW files contain no built-in sharpening, color correction, cropping, or any other custom processing. This work must be done later in the computerized digital darkroom, using programs like Photoshop, Elements, Lightroom, and Aperture.

Lossless files in the worlds of audio and photography are by their nature very large so they can capture all the data. An original RAW photo from a Nikon D800 camera is gigantic — 50 to 70 MB in size! *A professional audio engineer's high resolution original files* also take up tremendous file space: a five minute stereo 2496 recording[1] requires 164 MB! An hour-long recording of a full orchestra on 48 tracks requires 46 GB of space! Similarly, a sports photographer who shot 1000 RAW still photos would need a comparable amount of storage space.

Whether your lossless files are audio or photographic files, the artist must still deliver a master after processing the files. A multi-layered Photoshop document made from a RAW digital "negative" can take up 600 MB or more, so the premasters used in audio and still photography face similar transmission obstacles. FedEx is in business because the data size of our raw materials means we must still send hard disks by physical means.

Respect the Data

Although premaster files are huge, the final masters must travel via the Internet, and this usually means using a lower-resolution, lossy format. For audio, this means AAC or mp3 files; for photography, *JPEG files*. A JPEG can save as much as 90% of file space if set to its "minimum" quality, but this will invariably cause visible degradation of the image. The same thing applies to audio files if too much of the original data is thrown away. Whether the medium is photography or audio recording, you must *respect the data* once it has been converted to a lossy medium. For photographers, that means never process a jpeg: don't color-correct it or increase its size. If you do, the artifacts will become obvious. Similarly, in audio, the final lossy AAC files still sound quite good if produced with good methods from high resolution sources, but they should never be further processed. The AAC is considered a final product.

Viewing or Auditioning Original Files

Another interesting analogy between audio and photography is that both need the special software and hardware. A RAW photograph can only be viewed with the kind of special software we mentioned earlier, but a calibrated video monitor is just as important to properly edit the files. High resolution audio files also require special software and hardware so they can be properly processed. Another important factor is the space where the files will be evaluated. High-quality photography or HD video should be viewed under controlled lighting conditions. When you look at an HDTV in the showroom under bright ambient light, the picture would look washed out unless the monitor were intentionally "hyped," i.e., made brighter and thus inaccurate. When you take the hyped monitor home, it will look bright and glary. Similarly, a piece of music processed specifically to compensate for small computer speakers or iPods will sound fatiguing, harsh, and small when auditioned on monitor speakers in an acoustically-treated listening room. The quiet venue required for critical audio listening is analogous to the darkened room needed to view photographs on screen. The noisy venue is analogous to the fluorescent-lit HDTV showroom.

Mastered for iTunes

It is widely accepted that AAC produces better sound quality than mp3, given the same bitrate. In addition, **iTunes Plus**, Apple's improved AAC format used by the Mastered for iTunes Program, uses a higher 256 kbps bitrate. This produces less distortion than the old 128 kbps rate. Some claim that 256 kbps AAC sounds equal to or better than 320 kbps mp3, and is therefore closer to the sound quality of a 16-bit audio CD or a high-resolution master.

Why High Resolution Sources are Better for Encoding

Up until February 2012, when the Mastered for iTunes Program was announced, the only format Apple would accept for encoding was 16-bit/44.1 kHz audio derived from CDs or individual 1644 audio files. But since then, iTunes has begun to accept higher resolution sources, which potentially can produce sonically better results, provided you and your mastering engineer pay attention to the guidelines in this book. As I mentioned in the Introduction, only major labels will initially be allowed to submit higher resolution files, but you should follow as many of these guidelines as possible, and archive your files in high resolution, because it won't be long before independents can participate in the Mastered for iTunes program.

There are two reasons why we get better sound when AAC files are derived from high resolution sources: It's not widely known that perceptual coding can capture and reproduce the dynamic range of a 24-bit input, which has a greater dynamic range than 16-bit PCM! On the other hand, AAC is not as quiet as 24-bit PCM because of its distortion and noise modulation. The second reason is the classic principle of *garbage in = garbage out*, or its converse: *a better source yields better results.*

Working in high resolution yields cleaner, lower-distortion sources, and since distortion accumulates, the result sounds better. You'll be glad you saved those high-res files!

Closer to the CD or the High Resolution Master?

Mastering engineers involved in the Mastered for iTunes program have confirmed the improvement in quality of iTunes files that are carefully produced from high-resolution masters. When making your own judgments, do not compare a pressed CD to an older iTunes encode or a new (Mastered for iTunes) encode. That's because a pressed CD is not a master: it is an end-product that may have come from a different master. It also includes 16-bit dithering that is not on the iTunes Plus master (which can affect its tonality and depth), and it may have higher levels or even clipping that are not on the iTunes Plus master (more about levels in Chapter 4, *Keeping it On The Level*). The proper way to judge improvements attributable to the Mastered for iTunes Program is to compare the original high resolution master to the CD and to the encoded iTunes file at matched loudness. If the mastering engineer makes all the optimizations discussed in this book, you may even discover that the iTunes file sounds closer to the high resolution master, and better than the CD itself! Apple provides tools to ease the process of comparison and sound optimization. We'll discuss this in Chapter 7, *Tools of the Trade*.

MP3s, Streaming, and Other Venues

Whether you are producing files for iTunes, mp3, or streaming, it's a universal principle that you will get better sound quality if you maintain high resolution until the very last step before encoding to any lossy format. Encode mp3s from the 32-bit float premaster file. Send the streaming service the highest resolution file they will accept. If the streaming service requires a 1644 file, adjust the level first before dithering to 16-bit to reduce or eliminate encode clipping (described in Chapter 4, *Keeping It On The Level*).

Transcoding and the "Space Monkeys"

Down-converting Linear PCM files to a lower resolution is a somewhat lossy process, but cross-converting from one lossy-coded format to another is far more damaging. Surprisingly, more than one client has asked me to master from an mp3 file! Often these requests

come from producers who send the demo mixes from the mix engineers by mistake. But sometimes a producer has lost the original sources and asks me to master from an mp3. Coded files are very fragile; they suffer from any kind of additional processing. After you master from an mp3, even though you deliver a linear (lossless) 1644 file, it has not lost the signature of its origin; when it is once again converted to mp3 or AAC it will sound terrible. If you're familiar with the "space monkey" under-water-like sounds your cellphone sometimes makes, you'll recognize the kinds of artifacts found in transcoded product.[2] Avoid transcoding at all costs. In photography, this is analogous to the JPEG discussion earlier in this chapter: if you take a low-quality JPEG photograph, process it in Photoshop or other specialty software, and then output it once again to JPEG, you multiply the visual artifacts. In motion video, an example of transcoding would be to take a DVD (which is in a lossy video format), edit it in an editor, add special effects, and convert it back to the lossy DVD or mpeg video formats. This will cause pixelation and other visual artifacts, the visual analog of audio space monkeys.

How to Get the Best Performance from an AAC codec

The iTunes' Plus lossy AAC codec has a pretty big "bit budget": up to 256 kbits can be allocated per second of digital audio, even more for short periods, since it's a variable bitrate. But this is still a finite "bit budget": the codec must still allocate these bits among all the frequency bands of the audio spectrum. If at any time too many of the bands are at a very high level, the codec can run out of available bits and produce sonic artifacts or distortion. Thus it can't do a proper encode job. However, if the encoder detects audio activity in some frequency bands, and little activity in others, it can put more of its resources into the active band(s), and be frugal about the bits allocated to the less active band(s). This means that a lossy codec can do a better job when working with natural-sounding audio that has good dynamics (and that does not fill all the bands all the time), than working with highly-compressed or processed audio. When the bitrate is higher, the codec has an easier time, but heavy audio compression can still cause it to choke up, as revealed with a special test signal.

Measurements of Codec Performance

The measurement of a special multitone test signal shown here was taken using a software application called Spectrafoo. The signal

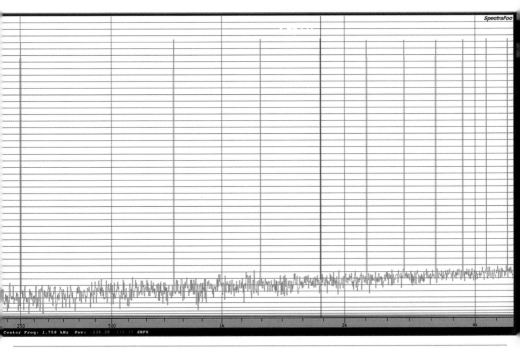

250   500   1k   2k   4k
Center Freq: 1.750 kHz Pwr: -135.35 ⸻⸻ dBFS

Original test signal (excerpt of the entire frequency range shown). Each frequency component is pure, shown as a straight vertical line, with a distortion-free noise floor. 500 Hz and 1 kHz tones are intentionally not included so that distortion in these regions will be clearly shown should they occur.

simulates music with a wide range of frequencies, no distortion, and a very low noise floor. Several carefully-chosen sine-wave frequencies have been distributed so as to expose a codec's weaknesses, purposely skipping 500 Hz and 1 kHz so the tester can look for distortion artifacts in those frequency ranges. Our thanks go to James Johnston (JJ to his friends), one of the world's experts on codecs and loudness, for providing this signal.

Next (pictured on page 48) we feed this signal to a low-bit rate AAC encoder (64 kbps) to set a "low-end" performance benchmark. I adjusted the gain to put its peak level just below the point where clipping would occur. This signal's BS.1770-2 loudness measures a very high -5.4 LUFS (Loudness Units Below Full Scale). I describe these loudness units in more detail in Chapter 5, *Loudness Normalization.*

As you can see, there's a lot of new activity around and between the original test tones, all created by the codec: significant signal-related noise and distortion have been added. There are many extra tones and noise where there was no original signal, and the noise and distortion floor rises where there is signal, producing noise modulation.

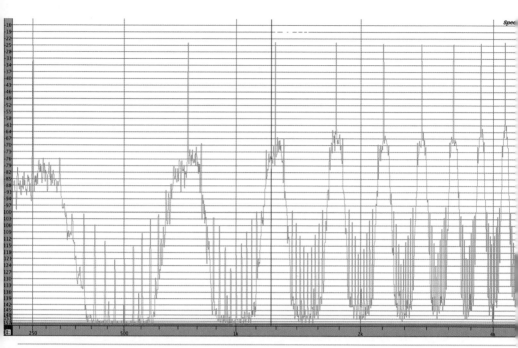

Test signal passed through a 64 kbps AAC codec (just below clipping). Note the extreme rise in the noise and distortion floor and the new discrete distortion tones in the spectrum.

Let's see what this test signal produces with an iTunes plus codec (pictured on page 49) which operates at a much higher bit rate, averaging 256 kbps.

The measurement shows that when the loudness of the material is kept to -17 LUFS (a very conservative level, shown in black), the distortion is fairly low. The codec is doing its job, keeping distortion and noise low in the areas where there is no signal. The noise floor modulates, rising in the areas where there is signal, because that's where the codec allocates its active bits. At this signal level, the distortion components are probably low enough to be psychoacoustically masked by the signal (i.e., they fall below the ear's masking threshold and so become inaudible). But the codec's distortion rises significantly when the level is increased. When the level is raised to a very high -5.4 LUFS, about as loud as the most-distorted pop recordings (shown in green), it may be hard to imagine that all this extra distortion energy can be masked. When we raise the level only a bit more, notice that a very small amount of increase in signal level (the sliver of red seen at the top of the tones) creates a huge jump in noise level (also in red). This sobering situation is evidence of clipping happening at the decoder input or output. It measures +0.8 dB True Peak level, which means that any DAC would be brought into clipping (Chapter 4 defines True Peak in more detail).

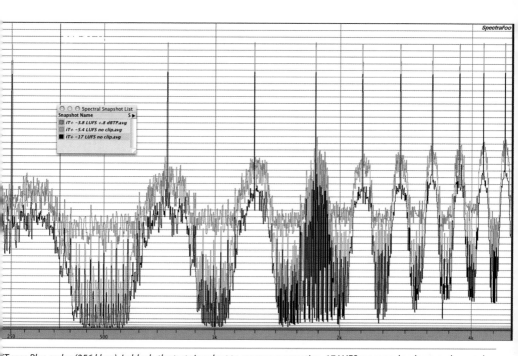

Spectral Snapshot List
Snapshot Name   5 ▶
☐ iT+ -3.8 LUFS +.8 dBTP.avg
☐ iT+ -5.4 LUFS no clip.avg
■ iT+ -17 LUFS no clip.avg

iTunes Plus codec (256 kbps). In black, the test signal set to a very conservative -17 LUFS program loudness and no peak clipping. When the level is raised to a very high loudness, -5.4 LUFS, just below clipping (green), notice how the distortion goes up significantly. One or two extremely-distorted pop recordings actually reach this level. When peak level is raised sufficiently to clip the codec (red), severe distortion artifacts are created which are quite audible.

These graphs demonstrate that lossy coding is a fragile technique and it must be carefully treated. On the plus side, lossy coding encourages a more dynamic approach to music and a return to sensible levels. My listening tests on the iTunes Plus codec show it to be quite transparent when fed conservative levels, with a subtle loss in depth and space that's virtually undetectable on most material. Be prepared to fail a blind listening test!

Lowering levels significantly would be a good idea, but few producers will tolerate their iTunes masters 10 dB lower than the competition! Chapter 5, *Loudness Normalization*, presents a hopeful situation, in which we will be able to lower levels so codecs do not distort as much. However, it is a very good idea to lower the level enough to keep the codec from clipping, as described in Chapter 7, *Tools of the Trade.*

Distortion is Cumulative

When mixing or mastering, consider the accumulation of distortion— "fuzz on top of fuzz," so to speak. Many of the processes engineers like to use, such as compression, limiting, tape saturation, clipping, spectral processors, tube units, and aural enhancers, need to be reconsidered if you're pushing the loudness, since the codec is going to saturate and contribute its own distortion. Should you lower the amount of processing, or switch to more-transparent processing, to help the translation to iTunes Plus? Not necessarily, but this is a consideration. In Chapter 7, I'll discuss some tools that Apple provides to make these decisions easier and simplify the job of mastering for iTunes.

---

1 A 2496 audio recording is technically not equivalent to camera RAW, but its resolution is close enough to the original resolution of the audio ADC for the purposes of our discussion.

2 Have you noticed that sound quality is considerably worse on a cell-to-cell call as opposed to cell-to-landline? That's because a cell-to-cell call is transcoded, but cell-to-landline involves only one conversion to the lossy coded format. The coded format used by cell phone calls is especially lossy, designed to capture voice through a vocal-modeling technique that barely provides good articulation with one pass through the codec.

"Do NOT Transcode. Never. Ever!"
Jim Johnston

O

2496   The abbreviation used throughout this book for 24-bit, 96 kHz audio.

128 kbps, 192 kbps, 256 kbps, 320 kbps   These are the most common bit rates (speed at which bits can be played back) used in lossy-coded audio, such as mp3 or AAC. All other things being equal, the faster the bit rate, the greater the resolution or signal-to-noise ratio of the lossy system. As a comparison, a lossless 1644 stereo PCM recording has a bit rate of 1.4 Mbps, more than four times the rate of 320 kbps lossy coding. Considering that, it's amazing how much audio information is retained in lossy-coded audio.

Bit rate   The speed at which bits are reproduced in digital audio. See kbps, see 128.

B

K

kbps   Thousand bits per second. The bit rate of a file. All other things being equal, the higher the bit rate, the better the sound quality of the file. But not all higher bit rates are created equal. For example, most authorities feel that an AAC (Advanced Audio Codec, also known as AAF (Advanced Audio Format) file at 256 kbps is equivalent to or better than the sound quality of an mp3 at 320 kbps.

Linear PCM (Linear Pulse Code Modulation)   The standard lossless method of encoding audio that gives each part of the dynamic range equal weight from the loudest to the softest sounds.

L

Lossless Coding   A method of coding audio that reduces its bitrate and file size without losing audible information. There is a limit to how low a bitrate can be achieved losslessly. Formats include FLAC, ALAC.

Lossy Coding   uses a perceptual (psychoacoustic) model to encode levels and throws out information based on the ear's inability to hear low level sounds in the presence of loud ones in the same frequency range. At low bit rates, lossy coding yields a compromise between sound quality and size of the file. From a distribution and speed of transmission standpoint, small size wins, but from an audiophile perspective, large size and high bit rate are preferred. Formats include mp3, AAC, AC3.

Loudness   The intensity of sound as judged and perceived by the ear, a perceptual quantity which can be estimated but never exactly measured. Try to avoid using the word "volume" in a scientific context when you mean loudness, since "volume" is technically the contents or size of a container.

LUFS   Loudness Units Below Full Scale. See Chapter 5, *Loudness Normalization.*

M

mp3   MPEG-1 or MPEG-2 Audio Layer III, by the Motion Picture Experts group. A method of lossy coding.

Transcoding   The process of converting from one lossy-coded format to another, e.g. from an mp3 to a cell-phone optimized format, or from a 256 kbps mp3 down to a 160 kbps. Transcoding is deleterious to the sound, because the sonic artifacts of each coding format or generation become added to each other. Broadcasters frequently meet this dilemma, as they are often asked to broadcast iTunes files, yet their broadcast format is usually lossy coded. Uh oh! Basic advice: Do not transcode!

T

True Peak (same as intersample peak)   Special "True Peak" meters have been developed that measure intersample peaks, which lets them anticipate the higher levels from DACs, filters, SRCs, and other processes. However, even the "True Peak" meter cannot estimate the effects of a codec until after decoding.

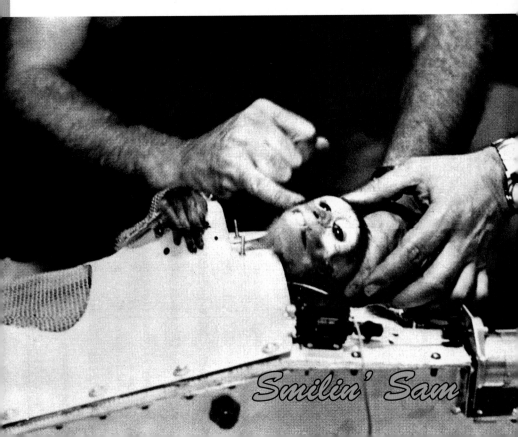

Smilin' Sam

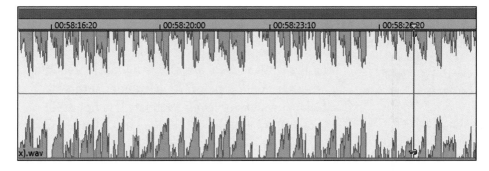

This mix received from client clips. But is it really clipped?

All the peaks (both positive and negative) have been clipped, and there is audible distortion when the file is auditioned. But when the gain is reduced, the peaks of the waveform are still intact and the distortion is gone!

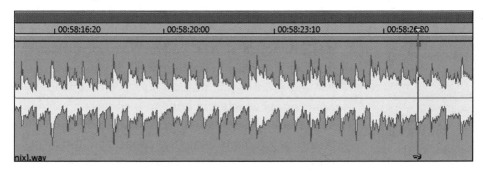

Same file after gain has been reduced, demonstrating that floating point rocks!

This apparent miracle is due to one property of the floating point file: *Floating point permits levels which are above 0 dBFS (full scale).* But remember that the real world is fixed point (all DACs require fixed-point signals), so unless the gain is reduced before playback, it will distort!

As long as all gain adjustment, EQ, compression or other processing is performed in floating point and has never been converted to fixed point, then levels can be above full scale (something which we have been taught never to do). A digital meter that is driven from the floating point circuit will show levels that are over full scale, e.g. +2.3 dB, in the red, which should alarm the operator. Eventually, levels that are over full scale must draw alarms, because we eventually have to convert to a fixed-point format, where we do the listening.

Perhaps the mix engineer didn't notice this distortion before passing the file on, or perhaps he performed some last minute off-line process without listening or generating a waveform. This "property" of floating

The Clipless Domain of Floating Point Notation

Floating point lives in a world of computation and can express signals which cannot exist in the real world. Fixed point files never exceed 0 dBFS sample peak, but floating point files may exceed 0 dBFS (the official full-scale point). This is legal only during computation so, before the file is converted to the real world of fixed point—the level must be dropped below 0 dB or the signal will clip or distort.

point even applies to sample rate conversion since most sample rate converters operate in floating point. It is possible to sample rate convert a 32-bit float file whose levels are above full scale, resulting in another floating point file at another sample rate, also with over levels, but which is not at all damaged! As long as you reduce the level before listening or converting to fixed point, the file is ok. Don't try this at home! Eventually you're going to get caught, so it always pays to practice safe levels. Meter your work and you won't get caught. I've received many Pro Tools Sessions whose mix bus was overloading simply because the operator did not insert a meter at the proper point in the master fader.

Be very cautious about levels during encode to AAC. Although all AAC codecs (to my knowledge) operate internally in floating point, if you feed a floating-point signal which would clip or a fixed point signal which is clipped to a codec—it will sound quite nasty coming out because the loud distortion products fill up the codec's limited bit budget. Even if the input signal is not clipped, codecs frequently produce a greater level on their output than input, so they can produce distorted AAC files that will clip when played back. As a rule, when coding, start with a signal which is not already distorted due to clipping, and if necessary, reduce the signal going into a codec until its output does not clip. In Chapter 7, *Tools of the Trade*, I'll illustrate some tools to facilitate that process.

Declipping Software

As long as we're on the subject of miracles, I'll bring up the **Declipper**, a fairly-new category of restoration software that is capable (to some extent) of "**undistorting**," literally restoring material which had been distorted due to clipping. This is performed by interpolation, which means that the declipper must guess at what got clipped, and if it guesses wrong, the cure can be worse than the disease. The first workable declipping software that I know of was invented by Dr. Andrew Moorer of Sonic Solutions in the mid '80s. It allowed you to identify individual spots that may have distortion due to overload and clean them up. Recent declippers allow you to define an entire region which has distortion which they will selectively correct. Declippers have been marketed by Cedar, Izotope, and others. Not all declipping software is created equal. In one case, I found the declipper introduced more audible distortion than it was trying to cure. In addition, a declipper is very delicate, requiring a careful and experienced operator with good monitors and good ears. But when it works, it's amazing to hear a purified sound and even restoration of the microdynamics. It stands to reason that the output level of a declipper will be higher than its input, so ordinarily you must turn down the gain to keep the restored peaks from clipping once again!

Sample Peak vs. True Peak

The maximum possible *sample peak*, or digital peak, of a recording is defined as 0 dBFS, the level that real-world digital audio samples cannot exceed. That's why we call it *full scale*. Sample peak, the actual peak numeric value of the samples, was the traditional method of peak metering until recently. But sample peak is not a very effective measure of judging overloads, because real world devices such as DACs, filtered processes such as SRCs (Sample Rate Converters), and any kind of lossy codec (e.g. MP3, AAC) *produce higher output level than their input*. Thus, a PCM signal can represent a signal that has higher amplitude (i.e. peaks) than the highest PCM (sample) value. A more effective measure of protecting from overloads, specified in BS.1770-2 is called *True Peak* (TP), achieved by upsampling methods. True Peak is a measure (actually an estimate) of intersample peaks (literally "peaks between the samples"), which will occur after some kinds of filtering. In practice this means that recordings whose sample peak is at or below full scale may overload after further conversion or encoding. These over-full-scale peaks are known as 0 dBFS+ levels. In the absence of a true peak meter, it is wise to keep sample peaks at or below -1 dBFS. If mastering studios ignore true peak they will soon discover that the PCM recordings that sounded so good in the studio become congested or distorted after conversion, especially to a lossy format.

A true peak meter does an effective job of predicting the output level of filtered processes such as DACs and SRCs, but lossy formats are a

Volume Control

Volume is measured in quarts, liters and cubic meters! In acoustics, the word *volume* is applied to the size of a room or enclosure. When I'm speaking colloquially with my colleagues, I may use the word *volume* or *volume control*, but when I need to be unambiguous I'll use one of the terms *level*, *gain*, or *loudness*. *Level* is an intensity or energy measurement, *gain* is the difference between two levels, and *loudness* is a perceptual quantity — what our ears perceive.

special case. Lossy formats not only filter but also add noise, which increases the sample peak level and the intersample peak. Because of this, a true peak meter cannot predict what a Codec will do, so be sure to measure the true peak level of the Codec (described in Chapter 7, *Tools of the Trade*). The lower the bitrate of the lossy medium, the higher the amount of overload distortion, so be mindful of those low-bit rate broadcasters, like satellite radio and some streaming services.

Clipping and the Loudness Race

By the year 2000, in order to "win" the loudness race for pop music, many mastering engineers began to clip the sound in addition to using hyper-compression and peak limiting. Clipping means to increase the level of the signal until it becomes squared off (clipped) at the maximum peak level. Its sample peak level cannot increase, but clipping does increase the average level, distortion and true peak (intersample peak) level. As I mentioned in Chapter 2, in an analog system, light clipping adds harmonically-related components to the sound, which can enrichen the sonics and sound musical to the ear. However, clipping in the digital domain produces harsh distortion that is not harmonically related, with substantial low-frequency grunge due to aliased harmonics. Sadly, many young people have become accustomed to the sound of extreme distortion and clipping because it's the only sound they know. They miss it when it's gone, partly because distortion increases the loudness (and louder is perceived as better).

Clipping at low sample rates is especially damaging because the alias frequencies are quite audible. An alias can be a difference or an additive tone. For example, 19 and 20 kHz may combine to produce a difference frequency of 1 kHz. This can only occur when there is a distortion mechanism in the audio system to create this alias or beat note.

Alias distortion can occur when levels would go over full scale, in digital compressors or other non-linear processors, as well as in lossy codecs. For example, the fifth harmonic of a 7 kHz signal, which is 35 kHz, beats against the 44.1 kHz sample frequency, producing an alias at 9.1 kHz (44.1-35). This will sound quite harsh. In simple terms, the higher the sample rate, the less annoying or destructive the alias distortion. In theory, an infinite sample rate is required to completely eliminate aliases for all harmonics of a source. But in practical terms, we must use sample rates from 4 to 8 times normal rates to produce a digital compressor or clipper that does not sound harsh. After processing, it downsamples back to the standard rate.

This image, courtesy of Jim Johnston, shows the terrible sonic cost of clipping a 21.533 kHz signal at 44.1 kHz sample rate. The horizontal axis of this plot shows the number of the FFT band, not the actual frequency, in a 4096 band FFT. Notice that while the original signal may be inaudible to most people, the additional aliased frequencies are going to be completely audible, inharmonic, and in fact they sound atrocious. Is it really possible that someone could get used to this sound?

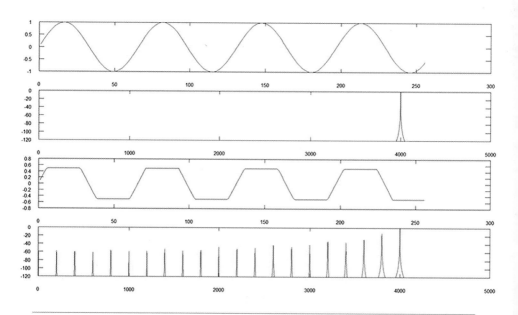

The cost of clipping in a 44.1 kHz digital system. The top image shows a 21.53 kHz sine wave. Second row shows its spectrum. Third row, sine wave clipped. Fourth row, the spectrum of the clipped result in a 44.1 kHz digital system. Courtesy of Jim Johnston.

Keeping It On the Level

C

Crest Factor   Formerly, this was the difference (in dB) between the highest peak level of a recording and its average level. But now that program loudness has been defined, Crest Factor becomes the difference between a recording's highest sample peak level and its average loudness, or the Peak-to-Loudness ratio of a recording.

Headroom   The potential difference between the loudness level of a recording and full scale sample peak. In other words, if the peak level of a recording does not hit full scale, its crest factor will be less than the headroom of the medium. In this book I use the convention of measuring headroom and crest factor up to the sample peak, but it would be legitimate (or perhaps preferable) to define it using true peak.

H

Hypercompression   An excessive amount of compression intended solely to make a recording sound louder, not necessarily for esthetic purposes. Of course esthetics are in the ear of the artist, and the language of sound expression is constantly changing. What was formerly considered "hyper" may now fall in the realm of "normal," for some people.

I

Intersample Peaks   Also known as True Peaks. Additional peaks between the samples that can occur when filtering, sample rate converting, or simply playing audio through a DAC. True peaks can have the same level as the sample peak or in practicality as much as 2 dB higher than the sample peak level in extreme cases. High level true peaks are quite common in highly processed work with a lot of compression, equalization, clipping, or peak limiting. The level of these peaks can be estimated with good accuracy by using an upsampling level meter, also known as a True Peak Meter. However, even the "True Peak" meter cannot predict the effects of a codec until the material has been encoded and decoded.

Macrodynamics/Microdynamics   Microdynamics refers to short term instantaneous or momentary changes in dynamics, such as during a snare drum hit or a sforzando. A recording may have a small macrodynamic range but still sound uncompressed with a good peak-to-average ratio if it has good microdynamics (if very little dynamics processing or compression was applied). For example, a Steely Dan recording sounds open, clear and pretty dynamic, even though it has a fairly small macrodynamic range, because it still has a good peak-to-average ratio with good transients, a natural, uncompressed quality. In comparison, a recording by the hard rock group, Tool, may have a large macrodynamic range (LRA), but compression processing has still been applied to reduce the transients when a powerful effect is desired. Such a hard rock recording may have a fairly low peak-to-average ratio, but still have a large LRA if the mixing or mastering engineer manipulates gain after the compressor. Such sonic differences are part of the engineer's style palette.

P **Partial Loudness** Loudness in one or more frequency bands, the sum of which makes up the total loudness.

Sample Peak The highest positive or negative value of an audio file, looking at the numeric value of the samples. See True Peak.

S

T **True Peak** (same as intersample peak) Special "True Peak" meters have been developed that (by upsampling) estimate the value of intersample peaks, which lets them anticipate the levels from DACs, filters, SRCs, and other processes which are often higher than the sample peak. However, even the "True Peak" meter cannot estimate the effects of a codec until after decoding.

Upsampling The same as oversampling for the practical purposes of this book. A processor which upsamples uses a sample rate converter as its first stage, converting the incoming rate to a higher rate than the source, usually 4 to 8 times the original rate. In a digital meter, this can be done for the purpose of capturing intersample peaks. In a compressor or peak limiter, upsampling can help produce less processing distortion. At the end of processing at the higher rate, the device then downsamples the audio back to the lower original rate on its output so it can be used in a plugin chain running at the original rate.

U

V **Volume** Volume is measured in liters, quarts, pints, gallons and cubic meters! Volume is a colloquial term often used for "loudness" but it is not an officially used term in physics or audio. Volume is also confusing and ambiguous as novice engineers confuse the *process* with the *result*, e.g.: "If I raise the volume (the process) do I get more volume (the result)?" So I avoid the term in print where possible. I occasionally use the term "volume control" because it does seem to be un-ambiguous, though will someone tell me what parameter "volume control" controls?

CHAPTER 5

Loudness Normalization

A Standardized Measure of Loudness

The loudness revolution has arrived, and is widely implemented throughout European broadcast networks and before the end of 2012, on U.S. television. This will eventually affect all recorded sound in specific ways. In Europe, both TV and radio sound levels are measured by the international ITU standard BS.1770-2, and are regulated by the European Broadcast Union (EBU) recommendation R128. In the U.S., levels are regulated by the Advanced Television Systems Committee (ATSC) A/85 specification. The European standard was adopted to obtain more consistent sound levels, and encourage better sound quality by requiring much less dynamics processing. In the U.S., however, the driving factor has been to eliminate loud television commercials via the CALM Act (passed by the U.S. Congress: Commercial Advertisement Loudness Mitigation Act). But the result is similar. It's a pity the CALM Act doesn't apply to U.S. radio, because this book focuses on making better-sounding music. I doubt that U.S. radio will voluntarily adhere to the CALM Act, since the forces of the loudness war only submit to official regulation.

Some of the measures defined by these standards appear on this next diagram of a recorded audio file with my additions.

Who can put the loudness genie back in the lamp? The answer can be found in this chapter.

The purple area represents the *loudness range* (LRA) of this recording, from the softest to the loudest passage, with its *average loudness* (also known as program loudness) in LUFS at the marked line. Its *crest factor* (peak-to-average ratio) is measured from its average loudness to the highest peak of the material (top of the yellow area). Its *headroom* (as I define it) is the maximum potential crest factor,

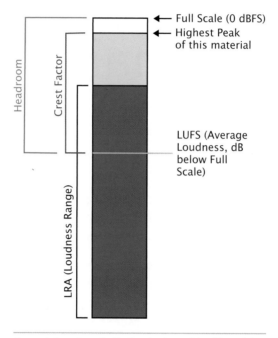

A recorded music audio file showing its peak level (top of the yellow bar), crest factor, loudness range (size of the purple bar), and the average loudness or program loudness (grey horizontal line within the purple area).

the distance between its average loudness and full scale (0 dBFS). For example, if a recording's average loudness is -23 LUFS and its maximum peak level is -3 dBFS, then it has a 20 dB crest factor and 23 dB headroom. Until now, headroom has been defined as the distance between the highest peak level and full scale (peak headroom) but now that the ITU has defined and standardized average loudness, it makes sense to define headroom in relation to loudness.

The ratio between the average and the peak level of a recording directly affects its sonic character, but the effect depends on the style of music. In percussive music, for example, having a relatively large distance between the average and peak level is important, because percussive music includes not only drums, but also the tap of a guitarist's hand on his guitar body for musical emphasis, the pizzicato clicks of a violin string and the attacks of a trumpet fanfare. But peak-to-average ratio is not that important in non-percussive music, which includes everything from string quartets to solo vocal recordings. Engineers manipulate peak-to-average ratio (crest factor) for special effect, by using compressors. Sampled drum sets used in hip hop are often highly compressed, with little peak energy above the average energy, so they can be made to sound very loud, and the subsequent distortion is part of the language associated with that musical genre.

Peak Normalization Exaggerates Loudness Differences

It is natural to have programs and musical styles with different peak-to-average ratios and different sound. The problem is that ever since the CD was invented circa 1980, engineers have been trying to peak all recordings to digital full scale, which causes extreme loudness differences between percussive and non-percussive recordings and between processed and unprocessed recordings. The following diagram compares the levels of a string quartet and a symphony orchestra, both adjusted until their peaks hit full scale, which is called *peak normalization.*

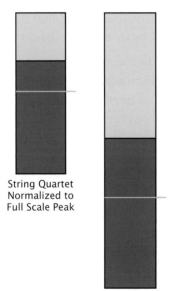

String Quartet
Normalized to
Full Scale Peak

Symphony
Orchestra
Normalized to
Full Scale Peak

The symphony orchestra needs much more peak headroom to accommodate all the momentary transients in the material, so peak normalization exaggerates the loudness difference. It makes the string quartet sound 10 to 14 dB louder than the symphony! This is especially problematic when constructing playlists in iTunes.

Peak normalization produces strange bedfellows, even in the classical music world. A peak-normalized string quartet sounds 10 to 14 dB louder than a symphony orchestra, making the listener run to his volume control every time he changes genres. This is a problem when constructing playlists in iTunes.

Peak Normalization Encourages Over-processing

Peak normalization also encourages program producers to compress or limit material in order to gain a loudness advantage; even classical music producers have asked me to peak-limit and raise the level of their program so it won't sound quieter than a competing product. It's very difficult to explain to a producer why a recording of string quartet accompanied by tambourine sounds quieter than a plain string quartet recording! But peak limiting can easily cause the sound quality to deteriorate; small amounts of peak limiting may be inaudible to some, but critical listeners will notice the deterioration of clarity, the increase in distortion, and the softening of transients. Unless we compress or limit, the presence of just one percussive piece on a classical album will require that the other tracks on the album be brought down to match its loudness which will, by necessity, be lower than that of another peak-normalized album that does not have any percussion.

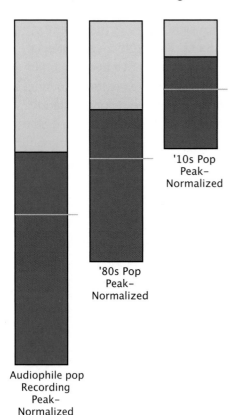

'10s Pop Peak–Normalized

'80s Pop Peak–Normalized

Audiophile pop Recording Peak–Normalized

In the same way, hard rock, metal, pop music and other heavily processed genres have suffered during the loudness race. Drums, particularly snare and bass drum in hip hop have lost their punch. There's just not enough room in the tiny yellow area to properly express these transients (illustrated at left).

Peak normalization has been the standard mastering approach since the invention of the Compact Disc in 1980. Because peak normalization makes compressed material sound louder, it has stimulated the digital loudness race.

Loudness Normalization

Radio and TV have always regulated sound levels to produce a more consistent consumer experience. They have also engaged in loudness wars to gain higher ratings. In the past (and continuing today in U.S. radio), broadcast loudness regulation was done by using severe processing and very strong compressors and limiters that squashed the sound, causing severe distortion. This distortion multiplies when already-distorted (hypercompressed) CDs are played on the radio. During the days of analog broadcasting there was probably no way to stop the practice of severe processing, but digital techniques permit all broadcast audio (except for live broadcasts) to be stored in computer files, which can be analyzed ahead of time and adjusted to a consistent loudness by simply adjusting their gain without further processing. Sonically, this offers a big advantage over the old way. Once the standards organizations defined a loudness measurement standard, and regulations came into place, true loudness normalization was under way in broadcast. For example, the classical and pop recordings described above can be loudness normalized by placing each of their average loudness points at the same "target level," shown here:

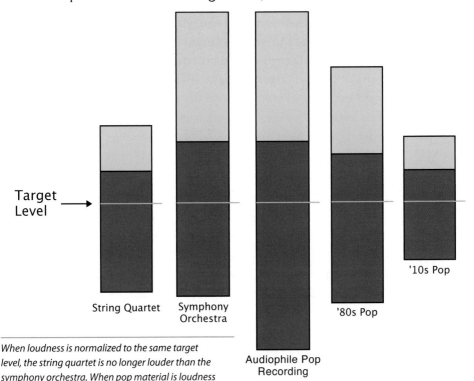

When loudness is normalized to the same target level, the string quartet is no longer louder than the symphony orchestra. When pop material is loudness normalized, there is no longer any loudness advantage to compressing or limiting the recording.

The Influence of Radio Broadcasting
on Music Production Techniques

Radio broadcasting has always had a profound influence on the sound of music and how we produce it. I believe the advent of loudness-normalized radio production in Europe is starting to have an effect on how we all produce and engineer music.

Consider this scenario: Whether you are producing popular or classical music, EBU digital broadcasts (and ATSC digital television broadcasts) normalize all recordings to -23 LUFS if they are not already at this level.[1] A well-managed U.S. TV or European digital broadcast network does not need to change the sound of your recording; they maintain its crest factor, since they use very little processing (no more loudness race). For example, let's imagine that our fictional country music song "You Break My Heart" was originally mastered with a crest factor of 11 dB (**Figure A** at right), and everyone is happy with the sound. One contributing factor is that the relatively high crest factor influences its clarity and impact. However, artist management decides it sounds too low compared to the competition. Keep in mind that management need not be concerned about how loud the song will sound on the radio, which broadcasts everything at the same average loudness. But they do want to impress radio Program Directors via loudness when Program Directors audition the CD for the first time. So they ask the mastering engineer to make it louder; he adds 2 dB of peak limiting, which raises the average level 2 dB and reduces its crest factor (**Figure B**). For better or worse, this becomes the release CD, not just for program directors to hear one time, but for posterity.

When this CD is broadcast on U.S. radio, its hot level interacts with the extreme processing, creating strong distortion and the song loses impact and clarity, especially on typical car speakers. But it sounds no worse than any other hot CD on popular U.S. stations. However, in European digital broadcasting, the song is not processed: the station simply lowers its average level to match their loudness standard of -23 LUFS (**Figure C**) and the sound is exactly like that of the release CD. Soon, the producers discover that on European radio, their song doesn't sound as good as some competing songs: it has less impact and is less clear on the radio than some of its competitors with a lower CD level. In response the producers release the original master as a new "radio version" (**Figure D**), which sounds much better on the digital radio: without the extra peak limiting there's more transient clarity, it's livelier,

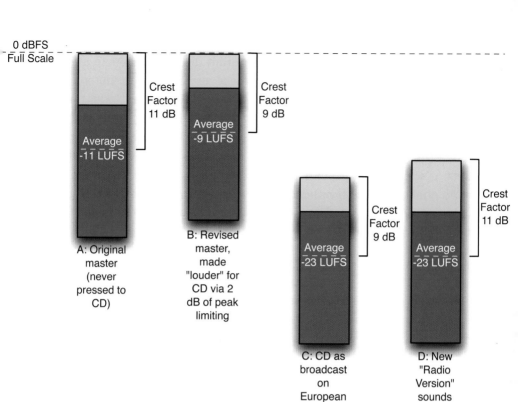

0 dBFS
Full Scale

Crest Factor 11 dB

Average -11 LUFS

A: Original master (never pressed to CD)

Crest Factor 9 dB

Average -9 LUFS

B: Revised master, made "louder" for CD via 2 dB of peak limiting

Crest Factor 9 dB

Average -23 LUFS

C: CD as broadcast on European Radio

Crest Factor 11 dB

Average -23 LUFS

D: New "Radio Version" sounds better with 2 dB less peak limiting

bigger and better than its competition! All this because European radio has not processed the sound, other than to normalize its loudness.

Note that examples A and D should sound the same if the playback level is adjusted to be the same, but example D shows that a loudness-normalized medium gives the producer the freedom to decide the sound quality he wants without causing an increased crest factor to run into the top of the medium. I'm not claiming that crest factor is the only, or even the most important, influence on the sound character of any recording. It's just that, in the ultimate days of the loudness race, crest factor has been reduced so much that it cannot help being a strong limiter (pun intended).

Incidentally, this new "radio version" also sounds better on U.S. radio, because it does not push the processors as far. Essentially, the producers have done an end run around the radio program director who picked their product when they gave him that hot CD. Now they can make it sound better on the radio!

This experience reveals one thorn in the side of the effort to end the loudness race: the manner in which program directors evaluate artists' CDs. They often put in one CD, listen for a very short time, put in another, with little attention to its musicality or dynamics. Under those conditions, there's no question they are influenced by the loudest disc, forgetting that the loudest disc sounds worse on the radio (it loses the race). If the disc is not loud enough to meet their preconceptions, they reject it, so the program director becomes a major influence on the sound quality of our recorded product. If program directors switch to evaluating music using iTunes Sound Check (see below), that will make all the difference in the world!

iTunes Sound Check Technology

A cross-genre iTunes Playlist exhibits big loudness jumps, especially between older and newer material. The solution for iTunes as for broadcast is **loudness normalization. Sound Check** is iTunes' loudness-normalization technology. It can be enabled in the iTunes preference settings. From then on it will work transparently and seamlessly.

Upon playback, the gain of each file is adjusted to eliminate sudden increases or decreases in track-to-track sound level as you listen. The music files themselves are not altered in any way by this process, which is reversible. With Sound Check activated, modern hip hop can play next to classic rock, and dynamic material can sit effectively next to compressed material in any tailor-made playlist.

Importing and Sound Check

iTunes stores its metadata in two possible places. AAC files downloaded from the iTunes Store incorporate a Sound Check metadata value in their headers, which will be used by Sound Check. AIFF files and standard WAVs (as opposed to BWFs, Broadcast WAV) do not incorporate loudness metadata in their headers so iTunes must keep this data in its database. It is possible that certain file formats imported under certain combinations may not have their metadata calculated, which has led some users to conclude Sound Check is broken. I assure you it is not broken, at least as long as you use AAC format throughout. Please visit our forum, where we will investigate the other permutations and file formats (e.g. AAC, AIFF, ALAC, mp3, WAV).

Sound Check: On By Default?

I believe Sound Check will be turned on by default in future versions of iTunes and Apple iOS devices, though Apple has made no announcements about this as of the publication of this book. Sound Check on by default will be a game-changer: although producers have no control over the playback loudness of their product (as if they ever did), they can create a new product with much more control over its clarity, microdynamics and spatial characteristics. The word *competition* takes on a whole new meaning: with loudness normalization, life in the music world just gets better. When producers listen to their songs on iTunes with Sound Check turned on, they discover that overprocessed material sounds worse—you can't fool Mother Nature! This is borne out in the European broadcast example shown above.

Nearly all music playback platforms are capable of loudness normalization: new cars come equipped with iPod jacks, and the iPod and iTunes are rapidly replacing the Compact Disc player in the car and at home. Other playback systems, such as Logitech's **Squeezebox**, and some digital music services, such as **Spotify**, also use loudness normalization technology, such as **ReplayGain**.

The Benefits of Sound Check

During the mastering session, producers can rest assured that Sound Check will maintain their album as loud as any of its competitors. Sound Check offers many advantages for both producers and consumers:

- Producers can make choices about how much distortion or dynamic manipulation they want to apply, without worrying that their recording will sound softer (or louder) than the competition

- The mastering engineer does not have to combat severe compression by over-equalizing

- Bass drums in hip hop recover their boom, punch and crack. Peak limiting and clipping return to optional creative tools, allowing drums to sound louder and more effective, since excessive processing can be avoided

- Snare drums and snappy instruments sound lively again

- The word "headroom" finally means something!

- Choruses sound louder than verses (once again)

- Space and depth are restored

- Music lovers can rediscover the sound of classic recordings, which no longer sound much too soft compared to contemporary pop recordings

Sound Check is like a country album played backwards: you get your wife back, your dog comes back to life, your pickup truck is repaired, and you're out of jail free!

Listeners play hip hop and metal loudly because that's how these genres are meant to be played. The temptation to over-process stems from the desire to make loud genres sound loudest. But over-compressing a modern hip hop or metal album to make it louder is just a band-aid.[2] As soon as Sound Check is engaged, severely compressed pop material is revealed as sounding flat and lifeless, and all the weaknesses of over-processing become exposed. Once the effects of hypercompression are uncovered, a lot of producers and engineers will be scrambling to make music sound good again, not just loud. The only way to restore dynamic excitement is by remixing and/or remastering. So to protect your precious catalogue, every time you have an album mastered, remember to produce and archive high-resolution dynamic masters that can benefit from iTunes Sound Check.

Sound Check's ecumenical nature can only equalize loudness, it cannot decide for you which genres should be played louder; but at least every genre is on a level playing field, with its sonic potential properly expressed, and as much esthetic compression or processing as the producer desires.

Everything Louder than Everything Else

The intent of loudness normalization systems is to regulate loudness for the consumer as well as eliminate a loudness war, but don't bother to ask a producer to voluntarily make his production sound equal to its competition if he can make it sound louder! Loudness normalization systems store the loudness information of a song in metadata, which, in Dolby's **Dialnorm** system, is in the hands of its producers. So any program producer can literally hack the metadata to make a song apparently louder. This includes discs encoded in Dolby Digital and Dolby True HD, resulting in a perversion of Dialnorm's entire intent: nearly all music DVDs and Blu-Ray discs have been given a Dialnorm value of -31 (the highest gain) by the program producers. For all of these discs, the playback system applies the maximum possible gain, which makes a mockery of loudness normalization. There's no turning back the clock once a single competing disc has been pressed and released with incorrect Dialnorm: it sets the war off and lets the loudness genie out of the bottle. For these reasons, many popular music video discs have been produced with as much distortion as the hottest pop CDs.

"Sound Check is like a country album played backwards: you get your wife back, your dog comes back to life, your pickup truck is repaired, and you're out of jail free!"

Sound Check Means No Cheating

There's only one way to put the genie back in the bottle: use a secure normalization system. This marks an important distinction between Apple's Sound Check and other systems: *Sound Check Means No Cheating.* Sound Check is a closed system, putting loudness normalization securely under the control of an independent third party. Apple is securely in control of the integrity of the audio from delivery of the masters. A unique vendor encodes all loudness metadata, protects that data, securely distributes, sells, and (in the case of streaming), broadcasts the material. Big Brother may be watching, but his intentions are benign and beneficial, protecting the best interests of all music producers who might otherwise forge loudness data to their advantage. And that's why Sound Check will succeed where other systems have failed.

Apple does provide tools for producers to evaluate Sound Check metadata in advance (see Chapter 7, *Tools of the Trade*). Engineers can make test encodes, see how the material sounds in comparison to the competition, and inspect the Sound Check values that will be given to their material. But this metadata cannot be submitted to Apple. They control the final encode and securely compute the loudness metadata. The Sound Check value Apple gets will be the same as the one you would get by importing your music into iTunes.

Noise is Not a Problem

With loudness normalization, system noise does not come up perceptibly even though the audio levels of some products have been

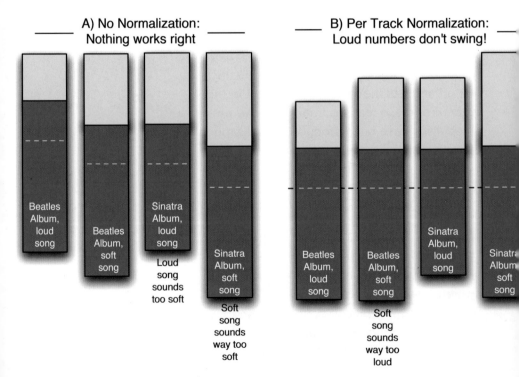

lowered. That's because listeners keep their volume controls at a reasonable, constant gain. It is a myth to say that we have to "use up all the bits," since peak bits have little or nothing to do with perceived loudness or signal-to-noise ratio.

Album Normalization: Soon!

Producers and mastering engineers adjust the levels of songs in an album according to how they feel, they do not regulate them to the same measured loudness. Adjusting all songs to the same loudness would be silly; ballads would be the same level as rockers, and the slow, relaxed movements of symphonies would be just as loud as the final, bombastic movement. Even when a soft movement from an album is placed into a playlist with loud material, it's usually better to respect the original relationship of the soft piece to the loud numbers. When Frank Sinatra sings a ballad, it's supposed to sound a little softer: he can mix quite well in the same playlist with rockers in an album-normalized system, as can be seen in this diagram.

The dashed line represents the average loudness of the song. In Figure A, we combine Frank with the Beatles without any normalization. Without normalization, the loudest song on the Sinatra album is as soft as the softest song on the Beatles album. Each album plays fine within itself, but either Frank has to be turned up or the Beatles have to be turned down, so this mixed genre playlist doesn't work at all.

C) Album Normalization: Everything feels right!

In Figure B, we perform **per-track normalization**, which adjusts each song to the same loudness (the Target Level Line). The soft songs sound way too loud. In fact

Mixed Genre Playlist. With album normalization, we can successfully combine a Frank Sinatra big band album with The Beatles.

the album sounds compressed because songs that are intended to be soft are made loud. Consequently, the loud numbers aren't swinging as they should compared to the soft numbers. Frank and the Beatles perform together, without any musicality. In Figure C, we perform **album normalization**, which adjusts the loudest song of each album to the target level. Now the loud songs of both performers are set the same, and the soft songs fall into blissful proportion. Sinatra and the Beatles can swing together!

Unfortunately, Sound Check does not currently support album normalization. It has normalized per-track since its inception. Apple has not announced plans to support album normalization, but I have no doubt it will be a feature in the future. There is only one virtue to per-track normalization: Dance Parties, where everything is supposed to be equal loudness. But in all other cases, album normalization must become the standard.

---

1 ATSC A/85 differs slightly from EBU R-128, but for all practical purposes, both normalize to the same target level, within 1 dB of each other.

2 BAND-AID® is a brand name of Johnson & Johnson!

Loudness Normalization

ATSC   Advanced Television Systems Committee (a U.S. organization). The ATSC loudness standard is ATSC A/85.

CALM Act   A U.S. Law, Commercial Advertisement Loudness Mitigation. Requires compliance by December 2012 for Television stations only. It does not apply to U.S. radio (Too bad!).

EBU   European Broadcast Union. EBU uses the ITU BS.1770 loudness measurement standard and has enhanced the ITU standard with EBU Recommendation R128 to define meter ballistics and other meter characteristics that use the loudness standard.

ITU   International Telecommunication Union. A United Nations agency for information and communication technologies. The standard for loudness measurement has been implemented by this organization.

Loudness Normalization   A method of adjusting gain of all files or recordings so that each recording is reproduced at the same perceived loudness.

Loudness Range (LRA)   The difference between the loudest and softest passages in a recording. The new loudness standards have defined Loudness Range as the long-term range of a recording, also referred to as a recording's macrodynamics. LRA ignores extremely soft passages such as soft introductions, silences, and fadeouts, using a statistical approach defined in EBU's R128 recommendation. LRA is specified in dB — the higher the LRA, the greater the loudness range.

LU   Loudness unit. A measurement of loudness using a weighting filter conforming with ITU BS.1770-2. The LU is a relative unit whose zero can be assigned to any convenient reference (e.g -23 LUFS can be assigned to 0 LU). LU differences are the same as dB differences. In other words, the difference between -3 LU and -2 LU is one dB. We can say that a program which measures -3 LU is 1 dB softer or, if you prefer, 1 LU softer than -2 LU.

LUFS   Loudness units below full scale. Average loudness of a recording, also known as Program Loudness (PL). One LU difference is the same as one dB difference. In other words, -23 LUFS is one dB softer than -22 LUFS.

Peak Normalization   The practice of adjusting the gain of a recording until the highest peak reaches full scale. This does not regulate loudness!

Peak-to-Average Ratio   Same as Crest Factor. See Chapter 4.

ReplayGain   An open-source loudness normalization technique used by many independent players.

Sound Check   A loudness normalization technique used by iTunes.

Target Level   The loudness level that a loudness normalizer is working to achieve. For example, a song whose average program level originally measures -16 LUFS will be attenuated 7 dB by the normalizer to meet the EBU target level of -23 LUFS.

CHAPTER 6

How Loud
Is Loud

In this chapter we'll take a look at the new loudness meters, how to use and interpret them, and their relevance to music production.

The revolution in loudness-normalized digital radio and TV has inspired a plethora of loudness meters, in every shape and size, to fit every taste. But because the ear/brain is complex, no meter can anticipate our ear's perceptions at every moment in time. For example, when I master an album that has a tune with a soft, delicate ending, the beginning of the next tune usually sounds too loud. But if I play the middle of the later tune for a moment, it doesn't sound too loud, and when I play its beginning again, it also sounds fine: I feel no need to turn it down. What's happening is that in general the ear does not react in absolutes, but it is very sensitive to contrast. And because the ear says so, the beginning of the following tune is too loud (regardless of what the meter says). One solution is to try to increase the space between the tunes. Another solution is to slightly turn down the introduction to the next tune, just enough to reduce the disturbance on the ear, making sure that I can lead the listener smoothly on the upward journey through the next tune.

Audio Meter Weaknesses

Because of the aural sensitivity to contrast, a sudden burst of loud sound in the midst of soft sounds has a much greater impact than something of equal intensity in the middle of other loud sounds (did that make you jump!). No meter can make this type of loudness judgment, or substitute for a set of trained audio engineer's ears. Also, meters do not take into account the listening sound pressure level. Our ears are not flat, and, depending on how loudly we are listening, meters may over- or underestimate our sensitivity to low frequency sounds.

Audio Meter Strengths

Despite these shortcomings, meters are absolutely necessary. Our ears can be temporarily reset by listening to a loud passage for a long time, but a meter can help us verify that the soft part of an album is not too soft. Meters are very good at detecting a "domino effect," wherein each tune becomes slightly louder than the previous one, and the ear gets accustomed to the escalation. We must learn to work with the audio meter, to take advantage of its strengths, and to understand its weaknesses. In general, audio metering has matured greatly: it is far more effective now than it has been since the dawn of audio recording.

A Conceptual Loudness Meter

The conceptual loudness meter shown here has no dimensions or decibel values, but it should be fairly clear what "in the red" means in the meter below — or is it?

Pianissimo	Piano	Mezzo Piano	Mezzo Forte	Forte	Double Forte	Triple Forte	Fortissimo
Very Soft	Soft	Medium Soft	Medium Loud	Louder		Very Loud	Loudest

A small musical group, such as a string quartet, cannot generate as much maximum sound pressure level as an orchestra (85 dB for a quartet versus 110 dB or more for the peak of a Mahler symphony with chorus). Surprisingly, the musical scores for a solo instrument (such as a guitar), a string quartet, and a full symphony orchestra all use the same loudness markings, even though the symphony can easily drown out the guitar.[1] But composers and performers understand that the loudness markings in a score are relative to the number and types of instruments involved, and the capability of the performers. In this conceptual world, large, loud ensembles would drive the meter into the red, and small ensembles would reach lower levels. Every recording would be made and reproduced according to the absolute performance SPL.

But absolute performance SPL is far too broad a range for the average listener. A strange thing happens when recordings of these different kinds of ensembles are typically reproduced: consumers commonly normalize them to a similar playback loudness, and so rock recordings, symphonies, string quartets, and solo instruments are typically played within 6 dB of each other. In other words, consumers compress their playback with their own volume controls. This is probably a wise musical decision; few of us own playback systems that can cleanly reproduce a symphony at its performance loudness, nor would we all enjoy or tolerate listening to a symphony in our living room at its full 110 dB capability.[2]

The range from the softest to loudest passage is often compressed by the broadcaster or the recording/mastering engineer. In a typical bedroom (for night-time television listening), listeners may prefer around 72 dB SPL for whispers, 74 dB for spoken word dialogue, 76 dB when the protagonist shouts, and when the bomb goes off, about 78 dB (my estimates). This is a very compressed range of only 6 dB (not counting ambience and musical score), since softer listening levels require greater compression to prevent soft material from being lost or misunderstood. Listeners can accept a wider range if they are very tolerant or listening at a higher average level (usually not late at night). That's how TV sound was processed for broadcast in the U.S. prior to the CALM Act.

In real life, a shout can be more than 10 dB louder than normal dialogue, and a car crash 20 dB! I hope the post-CALM-act U.S. TV broadcast range increases—maybe there will be closer to a 10 dB difference between a shout and a whisper. And we're going to need a nighttime (compressed) mode to reduce this range for soft playback, or listeners will not tolerate a large broadcast dynamic range, especially if one's partner is trying to get to sleep! Dynamic range compression of broadcasts is described in the ATSC A/85 specification but not covered in this book.

In the home theater or dedicated music listening room, the sound level is usually louder, and we tend to choose material that is less compressed—normal dialogue falls at around the 77 dB level, and the loudest sound effect or loudest musical passage can hit 90 dB or more, a wider range of 13 or more dB for the primary (foreground)

How Loud Is Loud

material. This seems substantial, yet it is actually compressed reproduction compared to a large movie theater! No matter where sound is
reproduced, sound range is usually compressed in comparison to real
life. The range would too great for most reproduction systems, envi-
ronments, and listeners. The sound levels mentioned above represent
average measured levels using a slow-reading sound pressure level
meter—the momentary (short term) peak level can reach 10 to 14
(rarely 20) dB above the average level.

Forte-based Real-time Meters

When I master any kind of music I use a *forte-based* meter, which
puts *loud* or *forte* passages at 0 dB. This is a picture of the VU meter,
which is still one of the most commonly used forte-based meters.

In the VU meter, forte (loud) passages are supposed to reach 0
VU, and the fortissimo (loudest) passages would reach as much as
+3 VU. Knowing what "forte" should sound like, I aim for 0 VU in loud
passages. These days, physical meters are in short supply, and the VU
design is showing its age. The biggest weakness is that the top 50%
of the scale covers only 6 dB of the range, making users think that
quiet music is quieter than it really is (less meter movement for the
same decibel change). Notice how the distance between +1 and +2
VU is greater than between -1 and -2 VU. Medium levels of -7 to -10,
which are acceptable, barely move the needle, which tempts the user
to overcompress. Strange, isn't it? But a computer-based meter with a
linear-decibel scale can be easily conceptualized and produced, like the
simple *forte meter* pictured on page 87. Notice how each decibel
change is expressed with the same "meter needle" motion.

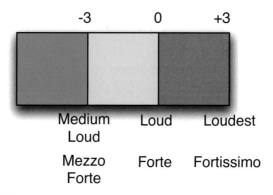

-3   0   +3

Medium   Loud   Loudest
Loud

Mezzo   Forte   Fortissimo
Forte

This simple forte meter has a linear-decibel scale, which means each decibel change covers the same physical distance. We've assigned 0 LU (loudness units) to forte, or loud.

The scale is defined in LU, *loudness units* relative to 0 LU, which we consider to be *forte*. Some rare recordings require occasional levels louder than fortissimo, which could be +5 or +6 on this scale. Loudness is an *average* (not peak) measurement, so peak levels can reach the top of the gray area shown below, which is 19 dB above the 0 LU level. I will show you where the +19 figure comes from in a moment.

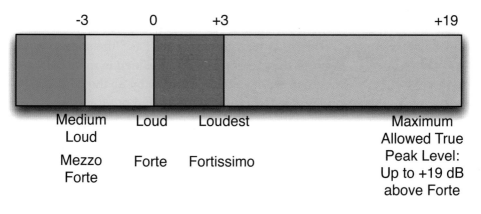

-3   0   +3   +19

Medium   Loud   Loudest   Maximum
Loud   Allowed True
Mezzo   Forte   Fortissimo   Peak Level:
Forte   Up to +19 dB
                                                      above Forte

This forte meter includes a grey area revealing that momentary peaks can occur up to 19 dB above the 0 LU (Forte) level.

A high amount of peak headroom allows music to breathe; forte and fortissimo sound louder and clearer in a medium that allows more peak level (headroom without limiting). However, in a highly compressed, peak-normalized pop music recording or medium, the gray bar can be very small, reaching no higher than +6 dB over this meter's 0 dB level. In that case, so much compression/limiting is needed to keep the peaks from overloading that the word *loudest* loses its meaning.

International Telecommunication Union, which has developed the BS.1770-2 loudness measurement standard.

EBU

European Broadcasting Union, which has elaborated on this standard to produce specifications for loudness metering and loudness normalization.

ATSC

Advanced Television Systems Committee, has adapted these standards for U.S. Broadcast.

Modern Loudness Meters Combine ITU and EBU Specifications

The serendipitous combination of ITU loudness standard BS.1770-2 and EBU recommendation R128/Tech Doc 3341 has produced the modern loudness meter, with at least five important advantages over the old VU meter:

1) Linear-decibel scales.

2) Equalized loudness units. Using the official unit of loudness, the *LU* (loudness unit), brings the meter's performance closer to the ear's perception of relative loudness, defined in the ITU standard. The ear is far more sensitive to high frequency sounds, so a high frequency sound with the same electrical energy as a low frequency sound should ideally read higher on a loudness meter. Consequently, loudness meters conforming to this standard are *equalized* (filtered) to approximate the ear's frequency response.

3) Standardized meter "ballistics," the speed at which the meter moves (as defined in the EBU documents), have been adopted by all the new loudness meter manufacturers and programmers.

4) For broadcast, a standardized average loudness of -23 LUFS, 23 loudness units below full scale. Later I'll discuss the relevance of the broadcast standard to music production.

5) Channel summing. From stereo to multichannel are summed into a single loudness measure.

EBU Loudness Scale

The conceptual forte meter on page 87 is based on a *relative* scale, relative to 0 LU, which represents *loud*. To determine the corresponding absolute digital levels for recording on the digital medium, we will look at the EBU and ATSC standards. In European broadcast, the EBU specifies

-23 LUFS as its standard average level, meaning that the average program must not exceed -23 LU below full scale digital. The U.S. ATSC standard specifies -24 LUFS average, but their tolerance is 2 dB as opposed to 1 for the EBU, so for all practical purposes the two loudness

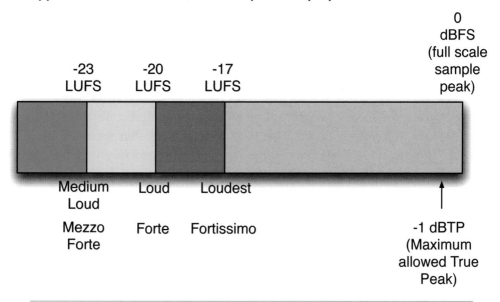

This meter superimposes the forte scale onto the EBU scale, using loudness units relative to full scale of 0 dBFS. When we correlate the EBU average level of -23 LUFS to mezzo-forte, forte lands at -20 LUFS.

standards are the same. Here is the EBU scale on an absolute basis, with the forte-based loudness scale superimposed.

EBU does not define *forte*: they simply specify an average loudness of -23 LUFS. But engineers like to have a real-time meter with a forte mark, so if we make the average loudness correspond with mezzo forte (medium loud), then approximately 3 dB louder would place forte at -20 LUFS. This worked out well when I produced an entire program metering with 0 LU set to -20 LUFS. Afterwards, the computed average of the program fell almost exactly at the required -23 LUFS. This means that an equal number of excursions went above mezzo-forte as below it. There can even be occasional loud excursions above the "loudest" mark with exceptional material, as long as the average does not exceed -23, so a program can still achieve a triple-forte climax if desired.

Although the EBU does not specify a maximum permissible loudness, they do require that true peaks not exceed -1 dBFS (or -1 dBTP). The permissible ATSC maximum true peak is -2 dBFS,[4] so the ATSC standard provides 1 dB less headroom if you use an average level of -23 LUFS. Since true peak levels nearly always exceed the sample

"These loudness meters are not just for broadcast!"

peak, sample peaks will never hit full scale. Finally, we have a metering standard that considers the output medium during production.[3]

Meters Meters Meters

The new styles of loudness meters fulfill every producer's need. There are real-time meters and software programs that can compute the average loudness of one or many files. In some meters, forte point can be set to 0, or any number the user prefers. Unlike broadcast, music recording has no standard average level. But I highly recommend taking advantage of the features of the new loudness meters. These new meters are not just for broadcast! Check out your recording's headroom, and your average loudness, which indicate the amount of compression. Mix engineers, set the meter to lots of headroom, and rely on your ears, or adjust the meter for the loudness goal you have set. Everyone should compare their program level against the approximately -16.5 LUFS target of Sound Check; learn the effect Sound Check has on your levels.

Vive La Peak Différence

One prime difference between most of these new loudness meters and previous types of meters is that the new meters *hide the real-time peak level*. This very important feature keeps operators from concentrating on the wrong detail: the peak level. The maximum true peak level is displayed as a warning only if it exceeds -1 dBTP or other user-set threshold. I do suggest that manufacturers add one more feature, which would let music-mixing and mastering engineers view the crest factor of percussive material, see how much of the available headroom they are using, and judge how much compression they are using. The feature would display crest factor without easily revealing the peak level: a dimensionless bar that can get taller from its center and change color as crest factor increases. The bigger the bar, the greater the crest factor.

A Mini-Gallery of Loudness Meters

Here's a "smorgasbord" of excellent loudness meters, from the fifty or more models that have appeared in the last few years! All of the meters and file measuring tools shown can display instantaneous or average (longterm) EBU or ATSC loudness, loudness range (LRA), and True Peak level. The realtime meters also display loudness history, each meter with its unique "sauce." You can find links to these and other manufacturer's sites at the digido links page mentioned in the Introduction.

Channel D AudioLeak

The AudioLeak (MacIntosh only) can run in real time, or as an extremely fast file analyzer or file normalizer with high quality dithering and sample rate conversion. I like the graphic display of audio level history, because it shows that high amounts of peak limiting (turquoise)

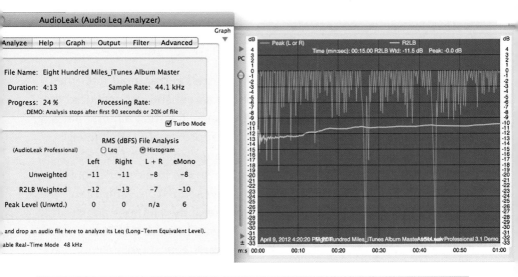

Channel D AudioLeak (offline or real time)

bring many of the peaks of the music to exactly full scale (which does not occur in real life)! The loudness line is displayed in yellow. The "Leak" name comes from "LEQ," another method of measuring loudness which the meter can measure.

Grimm LevelOne

The LevelOne is a cross-platform offline file-based metering and processing tool that meets EBU, ATSC or any user-configured loudness goal. The LevelOne's primary purpose is broadcast. It can output to a new file, adjusting (normalizing) material to the desired target level. Be aware that many broadcasters prefer to ingest material themselves

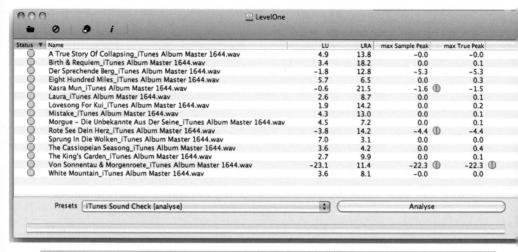

Status ▼	Name	LU	LRA	max Sample Peak	max True Peak
○	A True Story Of Collapsing_iTunes Album Master 1644.wav	4.9	13.8	-0.0	-0.0
○	Birth & Requiem_iTunes Album Master 1644.wav	3.4	18.2	0.0	0.1
○	Der Sprechende Berg_iTunes Album Master 1644.wav	-1.8	12.8	-5.3	-5.3
○	Eight Hundred Miles_iTunes Album Master 1644.wav	5.7	6.5	0.0	0.3
○	Kasra Mun_iTunes Album Master 1644.wav	-0.6	21.5	-1.6 ⓘ	-1.5
○	Laura_iTunes Album Master 1644.wav	2.6	8.7	0.0	0.1
○	Lovesong For Kui_iTunes Album Master 1644.wav	1.9	14.2	0.0	0.2
○	Mistake_iTunes Album Master 1644.wav	4.3	13.0	0.0	0.1
○	Morgue – Die Unbekannte Aus Der Seine_iTunes Album Master 1644.wav	4.5	7.2	0.0	0.1
○	Rote See Dein Herz_iTunes Album Master 1644.wav	-3.8	14.2	-4.4 ⓘ	-4.4
○	Sprung In Die Wolken_iTunes Album Master 1644.wav	7.0	3.1	0.0	0.0
○	The Cassiopeian Seasong_iTunes Album Master 1644.wav	3.6	4.2	0.0	0.4
○	The King's Garden_iTunes Album Master 1644.wav	2.7	9.9	0.0	0.1
○	Von Sonnentau & Morgenroete_iTunes Album Master 1644.wav	-23.1	11.4	-22.3 ⓘ	-22.3 ⓘ
○	White Mountain_iTunes Album Master 1644.wav	3.6	8.1	-0.0	0.0

Presets [iTunes Sound Check [analyse] ▼]   (Analyse)

Grimm LevelOne File Analysis Tool and Processor (offline)

and convert it to their standard level, file format and sample rate (they
may even use the LevelOne!). In the above screenshot, I configured the
LevelOne to evaluate the levels of all the tunes in a rock album I already
mastered, setting 0 LU to -16.5 LUFS, the approximate target level
of Sound Check. As we discovered in Chapter 5, Sound Check does
not currently perform album normalization, so the ironic workaround
is to turn off Sound Check. For example, the song "Von Sonnentau…"
is meant to be a very soft, ethereal piece, yet if the album is adjusted
on a per-song basis, it would have to be raised by 22.3 dB! This would
obviously destroy the producer's artistic intent. The current solution in
the Level One is to find the loudest song, in this case, "Eight Hundred
Miles," and turn down all the songs equally by the amount indicated
(5.7 dB). Then you will get an idea of how this album will compare with
other albums that have been loudness-normalized to a -16.5 target.

Grimm LevelView

This is a realtime loudness meter (plugin for Mac or PC, AU, RTAS
or VST) that simultaneously views the current loudness (integrated over
3 seconds), and recent loudness, with individual displays of the last 10,
30, 90, and 270 seconds! This is far less complicated than it sounds,
because the display gives the effect of a single "bending" meter needle.
In the following image, LevelView is configured for a user setting of 0
LU = -12 LUFS. Notice the rather small LRA of 3.4 dB for this material,
which is driving hard rock with nary a pause! Loudness histogram
(distribution) is displayed at the lower left.

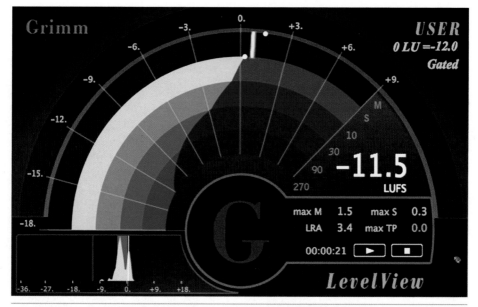

Grimm LevelView Loudness Meter (real time)

Nugen Audio VisLM-H Loudness meter

The Nugen Loudness Meter is a highly configurable cross-platform VST plugin with logging and a history display showing average, high and low loudness excursions.

Nugen Audio VisLM-H Loudness Meter (real time plus logging)

TC Electronic Radar Meter (LM5D)

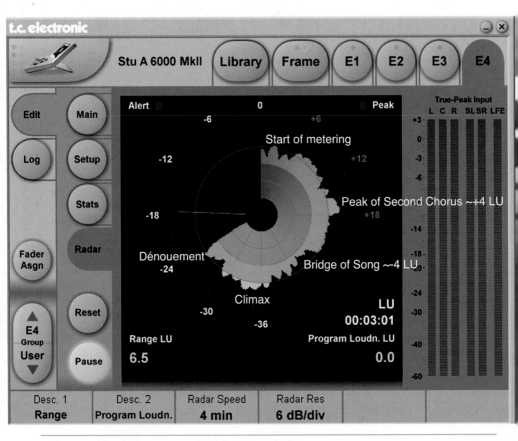

TC Electronic Radar Meter (real time, plus unique history display)

The TC Electronic Radar meter is available in three forms:

- A processing box popular with broadcasters with stereo digital (AES/EBU) input that connects to a PC via USB for its display

- A Mac or PC plugin for AU, VST, AAX and RTAS that displays stereo or 5.1 levels

- A stereo or 5.1 engine for TC's flagship System 6000 platform.

The meter is fully configurable: you can see from 1 minute to 24 hours worth on the "radar" display, or log loudness history to a text file. The structure of a song is easily identifiable; here we see the last 2½ minutes of a folk music production, beginning at "12 o'clock" and ending at "8 o'clock." The rings display a colorful level history. Each larger concentric ring of the radar represents an increment in loudness. For example, at 6 dB/division (ring), some of this musical material exceeds the 0 LU setting by as much as 4 dB (in yellow). I've noted significant parts of the song: notice how the bridge before the climax, the softest part of the song, leads to the climax. It's an excellent example of a dynamic folk recording, with about 8 dB range between loudest and softest foreground sections. The computed average loudness is 0 LU (which was set to be at a desired LUFS level) and its official loudness range is 6.5 dB (ignoring the softest and loudest passages). In this example, the display is paused so the real-time loudness level is not showing; it normally would appear in the outer spoked ring.

Conclusion

The new generation of loudness meters represents a major advance in the art of audio engineering. Thanks to clever new software interfaces, engineers are no longer restricted to looking at just the real-time audio level as they play a track: now they can see a rich "snapshot" of an entire audio track on-screen, in a single display. This reveals much more about the music than any VU meter ever could, and gives engineers a powerful new way to analyze audio. The result will be a better-sounding and more consistent product.

---

1 At the soft end of the musical scale, *ppp* is the abbreviation for *triple piano* or *pianissimo possibile* (meaning "softest possible" in Italian). *pp* is the abbreviation for *pianissimo* or very soft. *p* stands for *piano* or soft. At the loud end of the scale, *f* stands for *forte*, which means loud, *ff* for *double forte* (*fortissimo* or very loud), and *fff* means *fortissimo possibile* (the loudest possible), also known as *triple forte*.

2 I'm one of those audiophiles who does play symphonies and rock and roll at performance levels, but this requires a good listening room and the equipment to allow that.

3 There's more detail to the EBU R128 specification (at the digido links page), including discounting (gating) of soft passages and other features.

4 Leaving room for the higher peaks of lossy-coded Dolby Digital broadcast, which is common in the U.S.

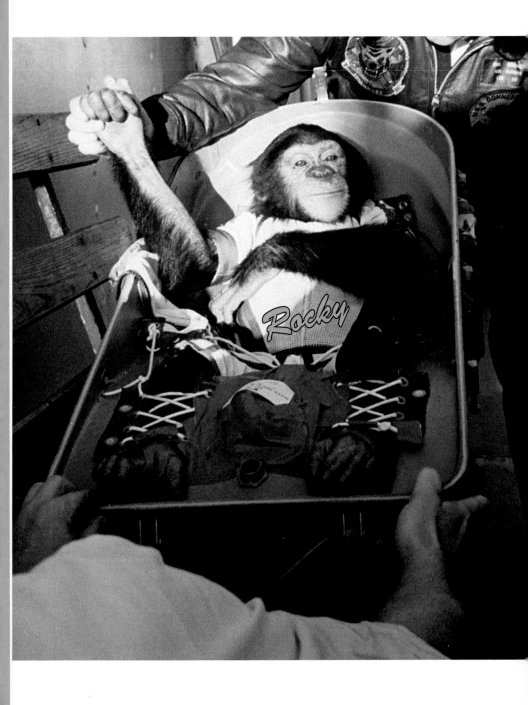

CHAPTER 7

Tools of the Trade

The Primary Tools of the Trade

In this chapter I'll describe some of the tools that we use during mastering and then introduce you to the Apple Mastering Tools designed to help improve the sound of coded product.

To start, the most important mastering tool we have is our ears, used within a proper monitoring environment. In my room, my clients and I can hear the effect of the smallest EQ or compression change, which helps us keep problems small enough so they fall "below the radar." A good monitor system helps us create a master that will *translate* to many other playback systems. A good monitor system also tells us when it's better not to process at all! Other primary mastering tools are the loudness meters and true peak meters described in Chapter 6.

The most used processing tool in mastering is the digital sample-accurate peak limiter, which has become so popular due to the loudness race. When overused, a limiter can reduce impact, depth and dimension, and add its own distortion. If the peak limiter goes into action only once or twice in an entire program and for a small amount of gain reduction, its presence can be invisible and it will have served its purpose of raising the average loudness to equal competitive product without hurting the sound. Interestingly, peak limiting is not necessary in a loudness normalized system with a low target level such as -20 LUFS, because there is plenty of system headroom. So the digital sample peak limiter could soon become an unnecessary tool, a thing of the past. But for today's world, I'm going to introduce a type of peak limiter that, when used subtly, can reduce distortion downstream.

Soft Wall or Brick Wall?

In the gentler days of analog, peak limiters played a different role, softening excessive transients or reducing "jumpy audio." Of course it's possible to design a digital limiter whose characteristics are as gentle as their analog brethren's, but then it couldn't eliminate short-duration over-levels. An analog limiter can subtly tame a snare drum hit that needs a little more "spanking" than a compressor can handle, but leave most of the transient and its sound character intact.

The higher the average program loudness, the more danger of peaks distorting and exceeding full scale, creating the need to control transient overloads with absolutely no overshoot, which led to "brick wall" digital limiters with instantaneous (zero sample) attack time. But fast attack times cause severe distortion and artifacts if the limiter is pushed beyond its "invisible" range. Depending on your opinion and the nature of the recording, the invisible range is from as little as 1 to as much as 6 dB. As the loudness race has progressed (if "progress" is the correct term here!), in competitive-level popular music, the peak limiter has become a nearly obligatory weapon.

	Attack Time	Ratio	Release time
Analog Limiter	10 to 30 ms or longer	10 to 20:1	100 ms or longer
Digital Sample-Accurate Peak limiter	Instantaneous	Infinite (1000:1 or >)	Program-controlled, between 10 and 100 ms. As fast as possible without revealing distortion

TC's Brickwall Limiter, which can upsample to prevent intersample peaks, as well as adapt its release time to reduce distortion.

Mastering engineers now have to decide whether to...

1) Clip the sound a little and peak-limit a little, which reduces the clamping effect of the limiter but replaces it with distortion

2) Drop the level, reduce or eliminate the peak limiting, strive for better sound when it's possible or permitted

3) Use the tools with grace, style, moderation and restraint

4) Give up mastering and grow asparagus (I've been tempted more than once!)

The good news is that the new breed of limiters can prevent inter-sample peak distortion and reduce its audible artifacts. They do this by oversampling the signal, which lets the limiter capture and control the level of the peaks that occur between the original samples. These peaks would otherwise distort devices further down the line — e.g. in DACs, sample rate converters, and broadcast chains. Oversampling has the side benefit of yielding reduced alias distortion.

Two excellent oversampled brickwall limiters are the TC Electronic Brickwall, pictured below (available in the TC Electronic Powercore or the System 6000) and the PSP Xenon (available as a multi-platform plugin). Both of these have adaptive release times. Release time defines how long a limiter takes to recover after it has brought down the gain of

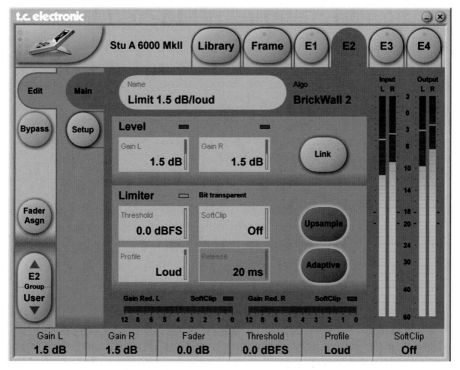

Tools of the Trade

a hot signal. A fast release time causes distortion, but if there is a very short (momentary) transient to control, a fast release time can be inaudible because the limiter gets out of the way before the ear can perceive the distortion. As a result, the sound appears louder because the gain is higher for a longer period of time. For the case of more continuous or longer sounds, a slow release time maintains low distortion. Every model of limiter has different algorithms for optimizing the tradeoff between invisibility, distortion, and apparent loudness. You can find other examples of oversampled peak limiters at our links page.

Clip It?

As I mentioned in Chapter 4, *Keeping it On The Level*, clipping creates distortion, but for better or worse, it has become a tool that mastering engineers can use to make the sound appear louder when a client is trying to "equal the competition in level." Only your ears can judge whether clipping distortion is acceptable or over the limit. Some types of music are very transparent and cannot tolerate any kind of clipping, which you'll hear plain as day. Other types of music crave distortion. Your experience and your client's desires will drive your judgment. In Chapter 3, *Lost and Found*, I demonstrated that clipping is especially problematic during conversion to the AAC medium, so always consider the final medium when mastering. Preventing clipping distortion (or at least informing you of its presence) is a primary focus of Apple's Audio Mastering Tools. Keep in mind that "cheating" by first clipping the sound, then dropping the level before encoding, is not a cure for bad-sounding AAC conversion, though it reduces the most egregious coding artifacts. Clipping at any point in the chain produces dense extra harmonics that cause a codec to overwork, and adds distortion.

Apple's Audio Mastering Tools

Apple developed these free tools to help content producers get the highest-resolution performance from the Mastered for iTunes program. These software tools help convert from high-resolution PCM to AAC, evaluate sound, compare an AAC encode against the original source, and reveal clips that can occur during an encode. Take advantage of these tools: they're your last line of defense.

The Apple tools are only available for MacIntosh computers, and can be installed in Mac OSX versions 10.6.8 (Snow Leopard) and up. Be aware that the AAC encoder in OSX Lion (10.7) has been improved, so you should at least be running Lion to hear AAC exactly as Apple will encode it. The tools are accompanied by a PDF readme with installation instructions, and some of the tools include help files that are accessible from the terminal. Adventurous engineers can even examine the source code of the applescripts. But for us humans, the most useful, quickest and easiest tool, my favorite, is the **AURoundTripAAC** plugin, shown below running in Audiofile Engineering's Wave Editor app.

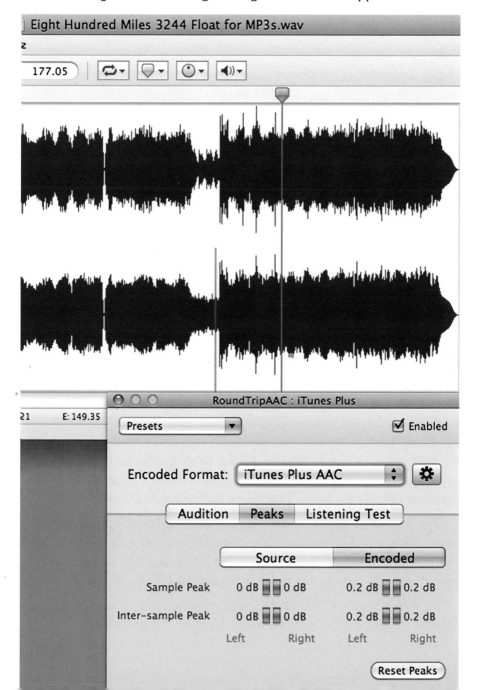

Apple's clever plugin allows real time auditioning of the codec and measurement of any clips (over levels) that may result from encoding. It takes the audio that the app is playing, feeds it to the codec, and then decodes it to linear PCM. The meters are fed from the floating-point output of the codec to display the amount of any over-levels on the codec's output. The plugin has a glitchless A/B switch to compare source vs. encoded, and a facility for a blind listening test to see if you can really hear the difference! The meters display both sample peak and intersample peak (true peak) level. It is the true peak level that you should be concerned about.

The workflow described in Chapter 2 began with a 96 kHz master file which was then converted to 3244 floating point in Weiss Saracon. By leaving it in floating point for as long as possible during the mastering process you avoid having to add multiple stages of 24-bit dither at each computation. Remaining in floating point, you also avoid any complications of overload or peak distortion should the SRC produce more level on its output than its input, the floating point file itself is not distorted. I do try to control the output gain of the SRC in advance during sample rate conversion so that it does not produce a true peak overload when converted to fixed point (linear PCM). But if an over is discovered, it is possible to take the gain down of the 3244 result without sonic cost as long as you do this before converting to fixed point or especially to AAC. This is fortunate because we expect the Apple Encoder to produce additional peaks, which will distort its AAC output if not attenuated beforehand.

That's why we have the Apple tools. I feed this 3244 "premaster" file through the Apple tool for auditioning and testing. If the codec produces overs, I can attenuate the 3244 file. I then apply 24-bit dither to create the 2444 master file to be sent to Apple. If, however, your source file is a 24-bit fixed-point file, you can attenuate it and add another layer of 24-bit dither (which will hopefully still be inconsequential). If your 24-bit source already contains overs, attenuating it may subtly help the encoding job, but it's better to go back to the drawing board.

The live output of this Apple plugin is delivered in floating point. It does not hurt (and it may help) the listening to insert a 24-bit dithering plugin at the end of this plugin chain. We'll have fun at the forum (see page 8) trying to prove this point and other related discoveries!

Let's take a look at the level measurements in the above figure. I started by playing the loudest material shown in the waveform. As you can see, the oversampling brickwall TC limiter was very effective, because intersample peaks (true peaks) of the source file were limited (i.e., kept down) to the same level as sample peaks (0 dBFS). However, the peak limiter was not able to anticipate the effect of the codec (only we can with this tool). The codec raised both the sample peak and the True Peak level to +0.2 dB. This can produce audible clipping artifacts (as we demonstrated in Chapter 3). After listening to the encoded material through the plugin, I felt that the clips did not sound good, and dropped the level 0.5 dB for good measure, as shown below.

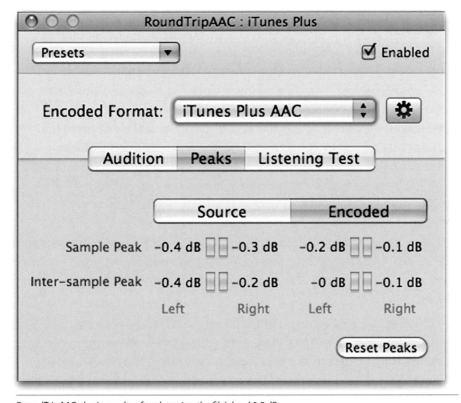

RoundTripAAC plugin results after dropping the file's level 0.5 dB

Notice that 0.5 dB of level drop barely squeaked by, as revealed by the 0 dB true peak level of the encoded output. When your source levels are scraping the top, AAC encoders are not entirely predictable; sometimes a given input level drop will not yield a proportional drop in the encoded product. Regardless, 0.5 dB drop produced no measured clips in the encoded material, though the level is just scraping the edge and the sound may be a bit gritty. Mastering engineers will advise that, if you've been pushing the distortion during mastering (perhaps clipping some internal stages), or using a lot of compression, it's a good idea to drop the level 1 dB or even more before encoding as a precaution if your client is interested in protecting sound quality. After finding the right gain for the 32-bit file, I attenuated it and dithered it to 24 bits, which then became the master submitted by the label to Apple's Mastered for iTunes (MFiT) initiative.

AAC is a very good-sounding codec, but it is not perfect. As you remember from Chapter 3, high average levels tend to saturate the encoder. Jim Johnston says that a perfect encoder should properly mask distortion from high average levels as long as it does not clip, but my listening tests show that reducing the level of even a moderately compressed acoustic recording by 10 dB produces a nicer-sounding AAC output, more open, with better transients. Reducing the amount of compression and therefore the average loudness also produces a better sounding AAC. Apple's encoder works at a variable bit rate so it can allocate more than 256 kbps when the task requires it. Maybe increasing the base bit rate is the key, but right now we've got an end medium that slightly softens the transients and reduces the stereo image a bit when it's pushed. The solution: *if it hurts the sound, don't push it!* Likewise, many mastering engineers are convinced that 1644 sounds smaller than 2444 so even CD format is less than perfect. Now is the time to stress that practices which produce a better-sounding AAC are also good for CD. Even though linear PCM doesn't saturate or distort like AAC, hypercompressed music still sounds bad on CD or any medium. Let's try to improve our sound practices for the benefit of all media.

I can use this Apple plugin in real time even though my mastering system is on a PC! I feed my signal out AES/EBU to a Mac running Digital Performer, which is set to input monitor and has the RoundTrip plugin engaged, and I can hear the effect of the codec in real time. For me, listening to the encode while mastering turns out to be overkill, as long as I make a 3244 (floating point) premaster, watch my levels with a true peak meter during mastering, and employ reasonably-conservative mastering techniques. The more aggressive the mastering, the brighter its EQ, and the more dynamics processing, the greater the chance it will overload any peak limiter, and certainly any codec. Sonnox/Fraunhofer makes a similar VST plugin for the PC but it currently does not support iTunes Plus AAC format.

To SRC or not to SRC?

Another great Apple tool is the **Master for iTunes Droplet**, a *drag and drop* applescript that takes an audio source file or group of files of any sample rate and produces iTunes Plus AAC files for proofing in iTunes. In addition, the droplet computes the Sound Check "profile" of each tune and embeds it into the metadata of the destination file, so you can hear the effect of the normalization when Sound Check is turned on. The droplet works with high resolution sources, performing all intermediate computations in 32-bit floating point until the final step. SRC, if needed, is carried out by Apple's mastering quality SRC. If you already have a mastering quality SRC (standalones such as Izotope, or Weiss's Saracon, or integrated in DAWs such as Pyramix or Wavelab) you can compare its quality to Apple's to decide if you should send 2496 or 2444 files to Apple for encoding.

Every down-conversion to 44.1 kHz is a slight sonic compromise, and not every style of SRC works best with all types of music. Several factors affect the sound of an SRC. The first is its distortion and low level resolution. Another is the effect of its low pass filter. If the filter is too gentle, the sound can alias. Material with muted trumpets or lots of high frequencies may need a steeper anti-alias filter to prevent distortion. However, it is commonly believed that a gentle filter sounds better to the ear. Visit the Infinite Wave site to view comparative measurements of SRCs, just remember that listening is as important as measurements.

Also note that Apple's tools are strictly for proofing: you must send standard WAV files to Apple for them to encode.

Apple makes another tool, available only from the command line in the terminal, called **afclip**. It can thoroughly examine an audio file, precisely identify where it clips, and create a graph of the exact times when the clips occur. This tool was developed before Apple perfected the round trip codec plugin with its own peak meter. I think that, unless you're trying to beat the reaper and confirm you have every tenth of a dB possible, you can make a good decision by spot-checking files using the RoundTrip plugin with a DAW that displays waveforms.

A Final Thought

This chapter completes our journey from the creation of digital music to the digital delivery to iTunes. I hope you've enjoyed *iTunes Music*. I have shared many of the techniques learned from working with high-resolution audio over the years so you can optimize the quality of your own iTunes music productions. The techniques described here will help you stay on top of this brave new audio world.

If you take home one lesson from these chapters, it's this: *Less is More.* Use less processing — you will have less cumulative distortion, which translates to a bigger, warmer, more attractive sound for the music lover. Processing in fewer steps at higher resolution means your audience gets to hear more from your production. Even if you are deliberately introducing distortion for an artistic effect, be aware that there are more steps in the chain beyond your listening point that can cause more distortion than you may want. Take a holistic approach for better results—that's what it's all about.

This is the last chapter in the book, but the conversation has just begun! Please join me online at the forum at digido.com (see page 8). I look forward to chatting with you and the many other readers of this book, as we explore the latest tips and techniques for producing iTunes files. Come discover how to get even more out of your music!

Rhesus

GLOSSARY

0

16-bit fixed   This is the standard fixed-point wordlength of the Compact Disc. 16 binary bits express a coded range of 96 dB, but with care and proper dithering, a 16-bit file can capture signals as low as -115 dB below full scale. The difference between floating point and fixed point notation is explained in more detail in Chapter 2.

1644   The abbreviation used throughout this book for 16-bit, 44.1 kHz audio.

24-bit   This is the fixed-point wordlength that modern ADCs are capable of capturing, which means they can code a signal as low as -144 dB below full scale, but with noise-floor considerations in real rooms and the thermal limitations of components, it's likely the lowest real-world signal able to be captured by a converter is -120 dBFS (dB below full scale), though some converter designers are claiming as low as -130 dBFS.

24-bit fixed   See 24-bit.

2444   The abbreviation used throughout this book for 24-bit, 44.1 kHz audio.

2496   The abbreviation used throughout this book for 24-bit, 96 kHz audio.

128 kbps, 192 kbps, 256 kbps, 320 kbps   These are the most common bit rates (speed at which bits can be played back) used in lossy-coded audio, such as mp3 or AAC. All other things being equal, the faster the bit rate, the greater the resolution or subjective signal-to-noise ratio of the lossy system. As a comparison, a lossless 1644 stereo PCM recording has a bit rate of 1.4 Mbps, more than four times the rate of 320 kbps lossy coding. Considering that, it's amazing how much audio information is retained in lossy-coded audio.

32-bit float   This is the most common floating point wordlength used for calculation by CPUs and DSPs. The difference between floating point and fixed point notation is explained in more detail in Chapter 2, *The Resolution Revolution*.

3296   The abbreviation used throughout this book for 32-bit float, 96 kHz audio.

44.1 kHz   The sample rate of the Compact Disc. This means that 44,100 samples of audio are captured in each second.

48 kHz   The most common sample rate used for digital video recordings, also used for original professional audio recordings. This means that 48,000 samples of audio are captured in each second.

96 kHz   A higher-resolution sample rate often used for high-fidelity music recordings. This means that 96,000 samples of audio are captured in each second.

A&R, A&R Director   Abbreviation for Artist and Repertoire. A title often given to the production director at a record label. His or her job is to work with and develop the artists signed to the label.

AAC   Advanced Audio Codec. Also abbreviated as AAF (Advanced Audio File format). Files which have been encoded as AAC may have one of these extensions: .m4a, .mp4, .3gp (the latter is used in cell phones).

ADC   Analog to Digital Converter.

AIFF, aiff   Audio Interchange File Format. The file extension is .aif. This was the file format most commonly used at Apple throughout the 80s and mid-90's but it has been largely replaced by WAV because WAV files (specifically Broadcast WAV files) can contain metadata while there is no metadata standard for aiff. Other than the internal numeric format, there is no audible difference between AIFF and WAV and the two formats can be interconverted with absolutely no loss.

Alias   See Alias Distortion.

Alias Distortion   An unwanted form of beat note, the result of interaction between the sampling frequency and an original signal (including the distortion components of the original signal), resulting in frequencies that are not present in the original signal. Any signal over half the sample frequency will yield alias distortion below half the sample frequency, and this distortion is most often neither harmonic nor musically consonant. For more information and an example see Chapter 4, *Keeping it On The Level*.

ATSC   Advanced Television Systems Committee (a U.S. organization)

Barcode Tips and Tricks   The barcode is a unique code that identifies product for sale. Normally, it identifies the entire album, but sales organizations have begun to use the album barcode as part of the product identification for selling singles derived from that album. But the album barcode is not enough: Each vendor or store must establish additional accounting codes besides the barcode to keep track of individual sales of singles from the album. The EAN barcode (International Article Number, formerly known as European Article Number) has 13 digits. The UPC (Universal Product Code, developed in the U.S.) has 12 digits. To maintain upward compatibility, older 12-digit U.S. UPC codes can automatically be converted to a legitimate 13-digit EAN code by adding an extra 0 at the left. However, all new 13-digit EAN codes used in the U.S. do not have to begin with 0! The checksum is the last digit to the right: it is a calculated digit used by barcode readers to protect against read errors. If you receive an 11-digit code from a distributor, it is a UPC without the checksum digit. 12-digit codes can be confusing. A 12-digit code could be a UPC number with a checksum, or the first 12 digits of an EAN without the checksum. Visit the digido links page described in the introduction and navigate to the barcode check site to help clarify the situation. The site can calculate a checksum digit that you can compare with the last digit of the number you were given. When in doubt, check with the supplier of the barcode.

Bit depth   See wordlength; they mean the same thing.

Bit rate   The speed at which bits are reproduced in digital audio. See kbps, see 192.

CALM Act   A U.S. Law, Commercial Advertisement Loudness Mitigation. Requires compliance by December 2012 for Television stations only. It does not apply to U.S. radio (Too bad!).

CD   Compact Disc.

CD text   Metadata which is embedded in a CD containing artist, title, genre and other information.

Checksum   See barcode.

Clipping   Digital Overload. The level of a digital signal cannot exceed full scale. When the gain would cause the signal to exceed digital full scale, the output becomes a squared-off wave, which looks "clipped," with moderate to severe distortion for the duration of the clipping. The output medium (e.g. DAC) can overload even if the sample peak of the file never reaches full scale. That's because many devices and processes produce higher output levels than their input, including filters, DACs, and codecs. So the sample peak level of the digital file should be somewhat below full scale to yield an output or playback that does exceed full scale (clip).

Codec   Coder-decoder. In general, the purpose of a codec is to lower the bitrate of the source audio so it can fit in a smaller space and be transmitted faster. Sometimes referred to as "data compressor," this term should be avoided, as it can be confused with audio dynamics compression. Just call it a "coder" or "codec" and the audio file "coded audio." It may sound strange at first, but it's the only unambiguous way to communicate. A lossy codec reduces the amount of information from the original source (via psychoacoustic means). A lossless codec does not reduce the amount of original information, so there is a limit to how low the bitrate can be reduced in a lossless codec. Examples of lossless codecs include ALAC (Apple Lossless) and FLAC (Free Lossless Audio Codec).

Compression   Reduction of dynamic range, reducing variations in level, which can be performed manually (e.g fader ride) or with a processor (compressor). Compression of macrodynamics is often performed by moving a fader up and down, while compression of microdynamics is done using a processor to reduce the short-term crest factor.

To avoid confusion, please do not use the word *compressed* to apply to coded audio. Reserve *compressed* to apply to dynamic range reduction.

CPU   Central Processing Unit. The core processor in a computer.

Crest Factor   Formerly, this was the difference (in dB) between the highest peak level of a recording and its average level. But now that program loudness has been defined, Crest Factor becomes the difference between a recording's highest sample peak level and its average loudness, or the Peak-to-Loudness ratio of a recording.

D/A   Digital to Analog Converter, DAC

DAC   Digital to Analog Converter

D/A/D   Digital to Analog to Digital. Serial conversion from one format to another and back again. For example, in order to process a digital signal with an analog equalizer, the signal must be converted to analog through a DAC (Digital to Analog Converter), then returned to analog through an ADC (Analog to Digital Converter).

DAW   Digital Audio Workstation. Some DAWs popular with mastering engineers (because of their unique features) include: Pyramix, SADiE, Sequoia, Soundblade, and Wavelab.

dBFS   Decibels below full scale. Often expressed with a space, e.g. dB FS. For example, -12 dBFS is 12 decibels below full scale digital.

dBTP   dB True Peak, referenced to 0 dBFS full scale. Often expressed with a space, e.g. dB TP. For example, a simple sine wave whose sample peak level reaches 0 dBFS will probably have a 0 dBTP level, but a more complex wave with lots of high frequency information or distortion can exceed 0 dBTP, e.g. +1.2 dBTP. See also True Peak in Chapter 5, *Loudness Normalization*.

DDP, DDP Fileset   Disc Description Protocol. An image file format developed by Doug Carson and Associates for use in manufacturing CDs, DVDs and certain other disc formats. It contains audio, track and metadata information, which the plant can use to cut a CD-Audio master for replication.

Dither   A special kind of noise which is used to reduce wordlength without adding distortion. See Chapter 2, *The Resolution Revolution*.

Double Sample Rate   Shorthand for audio files recorded at either 88.2 kHz or 96 kHz. Likewise, quadruple sample rate can be 176.4 kHz or 192 kHz.

DSP   Digital Signal Processor. Specially designed microprocessor that digitally affects the audio put through it. Depending on the algorithms designed for each specific chip some processes could include equalization, compression, limiting, bass management, etc.

Dynamic range   The difference between the softest and loudest signal, expressed in decibels (dB). Since a moment of silence before the music begins could be counted as contributing to a wide dynamic range, the official measurement unit of dynamic range is LRA. See Chapter 5, *Loudness Normalization*.

EAN   See barcode.

EBU scale   A loudness meter scale with equal "needle" excursion for each decibel. Generally, 0 LU is near the middle of the scale, with either +9 (normal) or +18 (extended) at the top. Different meter ballistics are available to suit either real-time or long-term judgment of loudness.

EBU   European Broadcast Union. EBU uses the ITU BS.1770 loudness measurement standard and has enhanced the ITU standard with EBU Recommendation R128 to define meter ballistics and other meter characteristics that use the loudness standard.

EDL   Edit Decision List. The layout section of a DAW where audio files may be placed, edited, faded, or digitally processed. EDL is a term used by SADiE. Pro Tools calls this a Playlist or Session; Sequoia calls it a VIP (Virtual Project); the principle is the same.

Fixed-point   A way to notate digital audio, with a fixed upper limit of 0 dBFS (full scale), and a fixed lower limit depending on the wordlength.

Flat   No equalization. The "EQ curve" for flat audio is a straight line.

Floating-point   A notation method for digital audio, with an extremely high upper and lower level limit. For example, 32-bit floating point digital audio can express level differences of thousands of dB, which is only useful in the virtual world of calculation, not in the real world of audio which has at most 130 to 140 dB of dynamic range, and in practicality only 50 or 60 dB.

Fraunhofer   A private company which invented the modern day lossless codec.

FTP   File Transfer Protocol. A standardized method for transferring large files that avoids using a web browser with its limitations.

Gracenote   A private company which maintains a database of artist, song and artwork information accessible by applications such as iTunes or Windows Media Player.

Headroom   The potential difference between the loudness level of a recording and full scale sample peak. In other words, if the peak level of a recording does not hit full scale, its crest factor will be less than the headroom of the medium. In this book I use the convention of measuring headroom and crest factor up to the sample peak, but it would be legitimate (or perhaps preferable) to define it using true peak.

High resolution   Audio which has a greater wordlength than 16-bits. A higher sample rate is also desirable. Having longer wordlength or higher sample rate does not necessarily specify a file's resolution, but it is a minimum requirement to be high resolution. See Resolution.

Hypercompression   Compression in an excessive amount intended strictly to make a recording sound louder, not necessarily for esthetic purposes. Of course esthetics are in the ear of the artist, and the language of sound expression is constantly changing. What was formerly considered "hyper" may now fall in the realm of "normal," for some listeners.

Intersample Peaks, True Peaks   Additional peaks between the samples that can occur when filtering, sample rate converting, or simply playing audio through a DAC. True peaks can have the same level as the sample peak or in practicality as much as 2 dB higher than the sample peak level in extreme cases. High level true peaks are quite common in highly processed work with a lot of compression, equalization, clipping, or peak limiting. The level of these peaks can be estimated with good accuracy by using an upsampling level meter, also known as a True Peak Meter. However, even the "True Peak" meter cannot predict the effects of a codec until the material has been encoded and decoded.

ISRC Code   International Standard Recording Code. A unique identifier for each song or version of a song. The ISRC code is part of the metadata. Each recording of a song gets a unique registered code that is used to track playback on the radio and keep track of royalties. While U.S. law does not currently allow for performance royalties, other countries do use the ISRC code to pay the performers. An ISRC code is supposed to remain as an identifier of a song throughout its lifetime, even if its owner changes (e.g. if one record company acquires the assets of another).

ITU   International Telecommunication Union. A United Nations agency for information and communication technologies. The standard for loudness measurement has been implemented by this organization.

iTunes   A cross-platform application developed by Apple Computer for playing audio and video. iTunes is now available on all Apple devices as well as Windows computers.

K

kbps   Thousand bits per second. The bit rate of a file. All other things being equal, the higher the bit rate, the better the sound quality of the file. But not all higher bit rates are created equal. For example, most authorities feel that an AAC (Advanced Audio Codec, also known as AAF (Advanced Audio Format) file at 256 kbps is equivalent to or better than the sound quality of an mp3 at 320 kbps.

LAME   A method of coding mp3 which produces slightly different audible results than the Fraunhofer codec.

LEQ   Loudness Equivalent. An earlier weighting method of measuring loudness which for broadcast purposes has been largely superseded by the ITU technique.

Level   A measure of intensity, voltage or energy, expressed in volts, dB, power or other units.

L

Linear PCM (Linear Pulse Code Modulation)   The standard lossless method of encoding audio that gives each part of the dynamic range equal weight from the loudest to the softest sounds.

Lossless Coding   A method of coding audio that reduces its bit rate and file size without losing audible information. There is a limit to how low a bit rate can be achieved losslessly. Formats include FLAC, ALAC.

Lossy Coding   uses a perceptual (psychoacoustic) model to encode levels and throws out information based on the ear's inability to hear low level sounds in the presence of loud ones in the same frequency range. At low bit rates, lossy coding yields a compromise between sound quality and size of the file. From a distribution and speed of transmission standpoint, small size wins, but from an audiophile perspective, large size and high bit rate are preferred. Formats include mp3, AAC, AC3.

Loudness   A perceptual quantity. The intensity of sound as judged and perceived by the ear. Try to avoid using the word "volume" in a scientific context when you mean loudness, since "volume" is technically the size of a container.

Loudness Normalization   A method of adjusting gain of all files or recordings so that each recording is reproduced at the same perceived loudness.

Loudness Range (LRA)   The difference between the loudest and softest passages in a recording. The EBU R128 recommendation has defined Loudness Range as the long-term range of a recording, also referred to as a recording's macrodynamics. LRA ignores extremely soft passages such as soft introductions, silences, and fadeouts, using a statistical approach defined in EBU's R128 recommendation. LRA is specified in dB — the higher the LRA, the greater the loudness range.

LRA   See Loudness Range.

LSB Least significant bit. The bit of an audio word that counts for the least amount of level. For example, in a 16-bit word, the MSB, or most significant bit, is at the left, and its state affects half the level of the word. The LSB on the other hand, affects as little as 0.001 dB of level.

LU Loudness unit. A measurement of loudness using a weighting filter conforming with ITU BS.1770-2. The LU is a relative unit whose zero can be assigned to any convenient reference (e.g -23 LUFS can be assigned to 0 LU). LU differences are the same as dB differences. In other words, the difference between -3 LU and -2 LU is one dB. We can say that a program which measures -3 LU is 1 dB softer or, if you prefer, 1 LU softer than -2 LU.

LUFS Average Loudness of a recording, also known as Program Loudness, measured in Loudness Units below Full Scale as per ITU standard BS.1770-2. One LU difference is equivalent to a 1 dB difference. For example, a program whose Loudness is -15 LUFS is said to be 5 dB louder than a program with -20 LUFS loudness.

Macrodynamics See Microdynamics.

Mastered for iTunes (Abbreviated MFiT in this book). A program or initiative created by Apple, Inc. in February 2012 that permits for the first time high resolution files to be sent to iTunes for encoding, and provides a set of recommended guidelines. See Chapter 7, *Tools of the Trade.*

Metadata Literally, "data about data." If we consider the sound of the audio or music to be its main data, then metadata is additional information about the audio file other than its audio. This extra data may be contained within the audio file's header (the so-called id3 tag) or in a separate database. Examples of metadata include album graphics, artist, title, composer, recording date, genre, and loudness information. In an audio CD, metadata is embedded in the subcode hidden within the disc's structure.

Microdynamics vs. macrodynamics Microdynamics refers to short term instantaneous or momentary changes in dynamics, such as during a snare drum hit or a sforzando. A recording may have a small macrodynamic range but still sound uncompressed with a good peak-to-average ratio if it has good microdynamics (if very little dynamics processing or compression was applied). For example, a Steely Dan recording sounds open, clear and pretty dynamic, even though it has a fairly small macrodynamic range, because it still has a good peak-to-average ratio with good transients, a natural, uncompressed quality. In comparison, a recording by the hard rock group, Tool, may have a large macrodynamic range (LRA), but still compression processing has been applied to reduce the transients when a powerful effect is desired. Such a hard rock recording may have a fairly low peak-to-average ratio, but still have a large LRA if the mixing or mastering engineer manipulates gain after the compressor. Such sonic differences are part of the engineer's style palette.

mp3 MPEG-1 or MPEG-2 Audio Layer III, by the Motion Picture Experts group. A method of lossy coding.

mp4 and m4a Apple uses mp4 as the extension for Video files encoded with AAC audio, and m4a for audio files encoded with AAC. The iPod accepts files with extensions mp4, m41, m4p (a protected AAC file), and the m4b extension for audiobook files (which can be either protected or unprotected).

MSB Most significant bit. See LSB.

Oversampling   See Upsampling.

Partial Loudness   Loudness in one or more frequency bands, the sum of which makes up the total loudness.

Pause Mark   In the Compact Disc, an optional subcode mark which defines the pause between tracks. However, audio can be included in the CD during the pause yet remain identified as "the pause," making it useful for a few features which iTunes did not inherit from the CD. See Chapter 1.

PCM   Pulse Code Modulation. The standard lossless method used to capture and reproduce digital audio. See Linear PCM.

Peak Normalization   The practice of adjusting the gain of a recording until the highest peak reaches full scale. This does not regulate loudness!

Peak-to-Average Ratio   See Crest Factor.

Playlist   An iTunes document that contains a list of songs in the order the user wants them to be played.

Producer   The individual(s) assigned to help create the sound of an audio production and work with the artist(s).

Program Loudness (PL)   See LUFS.

Quantization distortion   Distortion which can occur when converting between analog and digital or between two digital formats. This is not inevitable if dither is properly applied.

Radix point   In simple terms, this is the same function as "decimal point" for non-decimal notation. For example, in floating point, a dot, the radix point, is used to separate the exponent from the mantissa.

ReplayGain   An open-source loudness normalization technique used by many independent players.

Resolution   A rather loose term which in this book we use to define the wordlength and sample rate of a file. For purposes of definition, the lower the level of audio that we can "resolve," and/or the higher the highest audio frequency in the file, the better the resolution of the file. It's easy to say that a file is "high resolution," but a lot more difficult to prove it! If the file has been processed using high-quality techniques, then its resolution has likely not deteriorated significantly. The presence of high frequency information or of measurable low level information implies the file may be high resolution but not necessarily. Vice-versa, a recording with information "only" up to 15 kHz may be for many reasons a higher "resolution" recording than another which goes up higher. In the end, only the audio engineer knows the provenance of a complex chain of audio.

RMS   Root-Mean-Square   A means of calculating the average energy of a signal regardless of its waveform, in other words, despite the signal being time-varying, non-sinusoidal, or aperiodic.

Sample Peak   The highest positive or negative value of an audio file, looking at the numeric value of the samples. See True Peak.

Sample Rate Converter SRC, also known as **SFC** (sample frequency converter)   A device or software application that converts between one sample rate and another. Quantization distortion or loss of information can result if the SFC is not performed well.

Segue   From the Italian for "to follow," means to connect two songs together in an artistic way, usually by crossfading or blending the songs during the transition, sometimes by connecting the two songs instantaneously or abruptly.

Sequoia   A DAW produced by a software company called Magix.

SFC   See sample rate converter.

Single Sample Rate   Shorthand for audio files recorded at either 44.1 or 48 kHz. See double sample rate.

Sound Check   A loudness normalization technique used by iTunes.

SPL   Sound Pressure Level, the measured intensity of sound.

SRC   See Sample rate converter.

Subcode   Hidden code within the Compact Disc that contains information in addition to the audio.

Target Level   The loudness level that a loudness normalizer is working to achieve. For example, a song whose average program level originally measures -16 LUFS will be attenuated 7 dB by the normalizer to meet the EBU target level of -23 LUFS.

Track mark   A coded location in the CD subcode that indicates when a new track begins. There are no track marks in iTunes, the beginning of the audio file is the beginning of the track.

Transcoding   The process of converting from one lossy-coded format to another, e.g. from an mp3 to a cell-phone optimized format, or from a 256 kbps mp3 down to a 160 kbps. Transcoding is deleterious to the sound, because the sonic artifacts of each coding format or generation become added to each other. Broadcasters frequently meet this dilemma, as they are often asked to broadcast iTunes files, yet their broadcast format is usually lossy coded. Uh oh! Basic advice: Do not transcode!

True Peak   (same as intersample peak) Special "True Peak" meters have been developed that (by upsampling) estimate the value of intersample peaks, which lets them anticipate the levels from DACs, filters, SRCs, and other processes which are often higher than the sample peak. However, even the "True Peak" meter cannot estimate the effects of a codec until after decoding.

Glossary

UPC   See barcode.

Upsampling   The same as oversampling for the practical purposes of this book. A processor which upsamples uses a sample rate converter as its first stage, converting the incoming rate to a higher rate than the source, usually 4 to 8 times the original rate. In a digital meter, this can be done for the purpose of capturing intersample peaks. In a compressor or peak limiter, upsampling can help produce less processing distortion. At the end of processing at the higher rate, the device then downsamples the audio back to the lower original rate on its output so it can be used in a plugin chain running at the original rate.

VBR   Variable bit rate

VIP   See EDL.

Volume   Volume is measured in liters, quarts, pints, gallons and cubic meters! Volume is a colloquial term often used for "loudness" but it is not an officially used term in physics or audio. Volume is also confusing and ambiguous as novice engineers confuse the *process* with the *result*, e.g.: "If I raise the volume (the process) do I get more volume (the result)?" Thus I avoid the term in print where possible. I occasionally use the term "volume control" because it does seem to be unambiguous, though will someone tell me what parameter "volume control" controls?

WAV   Pronounced "wave." The file extension is .wav. The most commonly used source file format in digital audio (AAC is probably the most common file format overall). It is a lossless linear pcm format and may contain metadata in its header, the metadata standard defined in the Broadcast Wav specification.

Wordlength   The number of bits required to define an audio sample or "word." For example, 16 bits are required to define each CD word, the wordlength of the Compact Disc is 16-bit. Some engineers use the term "bit depth" to define the same meaning.

Bob Katz, Mastering Engineer

Bob Katz is one of the best-known mastering engineers in today's music industry. His book, *Mastering Audio*, has gone through six printings and two editions. For many mastering engineers, it is the bible. He has written for *dB* magazine, and was a columnist for *Resolution Magazine*. He has written many articles and reviews in publications such as *dB, RE/P, Mix, AudioMedia, JAES, PAR, and Stereophile*.

Bob's recordings have received disc of the month in *Stereophile* and other magazines numerous times. Reviews include: "best audiophile jazz album ever made" (McCoy Tyner: New York Reunion reviewed in *Stereophile*). "If you care about recorded sound as I do, you care about the engineers who get sound recorded right. Especially you appreciate a man like Bob Katz who captures jazz as it should be caught." (Bucky Pizzarelli, **My Blue Heaven**, reviewed in the *San Diego Voice & Viewpoint*). "Disc of the month. Performance 10, Sound 10" (David Chesky: New York Chorinhos, in *CD Review*). "The best modern-instrument orchestral recording I have heard, and I don't know of many that really come close." (Bob's remastering of **Dvorák: Symphony 9**, reviewed in *Stereophile*).

Bob mastered **Olga Viva, Viva Olga**, by the charismatic Olga Tañon, which received the Grammy for Best Merengue Album, 2000. **Portraits of Cuba**, by virtuoso Paquito D'Rivera, received the Grammy for Best Latin Jazz Performance, 1996. **The Words of Gandhi**, by Ben Kingsley, with music by Ravi Shankar, received the Grammy for Best Spoken Word, 1984. In 2001 and 2002, the

Parents' Choice Foundation bestowed its highest honor twice on albums Bob mastered, giving the Gold Award to children's CDs, **Ants In My Pants**, and **Old Mr. Mackle Hackle**, by inventive artist Gunnar Madsen. The Fox Family's album reached #1 on the Bluegrass charts. African drummer Babatunde Olatunji's **Love Drum Talk**, 1997, was Grammy-nominated.

He has mastered CDs for EMI, BMG, Fania, Virgin, Warner (WEA), Sony Music, Walt Disney, Boa, Arbors, Apple Jazz, Laser's Edge, Sage Arts and many others. Clients include a performance artist and poet from Iceland; several Celtic and rock groups from Spain; popular music groups of India; top rock groups from Mexico and New Zealand; progressive rock and fusion artists from North America, France, Switzerland, Sweden and Portugal; Latin-Jazz, Merengue and Salsa groups from the U.S., Cuba, and Puerto Rico; Samba/pop from Brazil; tango and pop music from Argentina and Colombia; classical/pop groups from China, and a Moroccan group called 'Mo' Rockin.'

Bob enjoys the Celtic music of Scotland, Ireland, Spain and North America, Latin and other world-music, Jazz, Folk, Bluegrass, Progressive Rock/Fusion, Classical, Alternative-Rock, and many other genres. For a comprehensive list of Bob's recordings and the great artists he has recorded or mastered, visit www.digido.com.

A Career in Audio

At age ten, Bob Katz began playing the B-flat clarinet and, between breaths, built and modified audio gear. He began his amateur recording career in high school, bolstered by his fascination with science and linguistics and his love of French and Spanish. Today, he gives seminars in three languages. Two years at Wesleyan University were followed by two more at the University of Hartford, studying Communications and Theatre and directing all recordings at the college radio station.

Bob taught himself analog and digital electronics through books, and learned audio electronics from several mentors and co-workers. Just out of college, he became Audio Supervisor of the Connecticut Public Television Network, producing every type of program from game shows to documentaries, music and sports, and he learned to mix all kinds of music live.

In 1977 Bob moved to New York City to begin a recording career in records, radio, TV, and film. At the same time, he was building and designing recording studios and custom recording equipment. From 1978-79, he taught at the Institute of Audio Research, supervised the rebuild of their audio console and studios, and began a friendship with IAR's founder, Al Grundy, a mentor and influence. Other New York era influences include designer Ray Rayburn and acousticians Francis Daniel and Doug Jones. In the 80s, one of Bob's clients was the spoken-word label, Caedmon Records, where he recorded actors such as Lillian Gish, Ben Kingsley, Lynn Redgrave, and Christopher Plummer.

A proud ex-member of the New York Audio Society, Bob was the ultimate audiophile, always improving the high-end stereo system he used to evaluate his and other's recordings. In 1988, he began a long association with Chesky Records, which became the premiere audiophile record label. In 1989 he built the first working model of the DBX/UltraAnalog 128x oversampling A/D converter, and produced the world's first oversampled commercial recordings. Bob recorded and mastered many albums for Chesky, including his second

Grammy-winner, and in 1997 the world's first commercial 96 kHz/24-bit audio DVD (on DVD-Video).

New Ideas For Recording and Mastering

For years, Bob specialized in minimalist miking techniques (no overdubs) for capturing jazz and other music that is commonly multimiked. Critics called his recordings musically balanced, exciting, and intimate, while retaining dynamics, depth and space.

In the 90s, Katz invented three commercial products found in mastering rooms around the world: The FCN-1 Format Converter, dubbed by Roger Nichols the "Swiss-Army knife of digital audio," his VSP model P and S Digital Audio Control Centers (Class A rated in Stereophile Magazine).

In 2006 Katz designed and patented a new category of product, the Ambience Recovery Processor, which uses psychoacoustics to extract and enhance the existing depth, space, imaging and definition of recordings. Weiss Audio of Switzerland, Algorithmix of Germany and UAD have licensed Bob's K-Stereo and K-Surround processes.

Digital Domain

In 1990, Bob founded **Digital Domain** and became a full-time Audio Mastering Engineer. There he applies his specialized techniques to bring the exciting sound qualities of live music to every form recorded today, mastering music from pop, rock, and rap to audiophile classical. In 1991, he began the popular digido.com website, the second audio URL to make the World Wide Web. You can read many of his articles and the comprehensive audio FAQ online.

Digital Domain provides services to independent labels and clients, including mixing, graphic design, replication and of course, mastering. Mary Kent is an accomplished photographer and graphic artist, the visual half of the Digital Domain team and more than two-thirds of the charm. She's author of the book **Salsa Talks!** Yes, she can dance. In 1996, Bob and his wife Mary moved the company from the Big Apple to Orlando.

Since this book is all about iTunes music, it is appropriate that Bob's and his wife Mary's lives are irrevocably intertwined with Apple products. Their first Mac computers were Bob's 1989 SE, soon upgraded to a IIfx and Sonic Solutions cards, and Mary's IICI, which she used for graphic design. Today, there are two Mac Pros, two iMacs, two Mac Minis, a MacBook Pro, a Time Capsule, an iPad tablet, an iPod touch, and two little iPhones scuttling underfoot.

Christopher Morgan, Editor

Christopher Morgan is a computer scientist active in the Boston-area hi-tech industry for over three decades. He has degrees in electrical engineering and computer science, and was Editor in Chief of *BYTE* magazine as well as Vice President of Lotus Development Corporation. He is an audiophile and a folk musician, and has been a member of the Audio Engineering Society (AES) and the Institute for Electrical and Electronics Engineers (IEEE). His books include *Wizards and Their Wonders,* about the top 100 people in the hi-tech field, and *The Computer Bowl Trivia Book.*

He is an organizer of the biennial "Gathering for Gardner" conference in Atlanta, where magicians, scientists, mathematicians, and artists gather to share the recreational side of mathematics, and has twice hosted the International Puzzle Party, an annual international convocation of puzzle designers and collectors. He is past president of the Ticknor Society, a Boston-based group of book lovers and collectors, and has served on the boards of the Boston Computer Museum, the Boston Museum of Science, and the Boston Lyric Opera. He is currently working on a book about Lewis Carroll's games and puzzles. Morgan is also a professional magician and photographer (www.morganpix.com).

JJ is presently an independent consultant, retired from the position of Chief Scientist at DTS, Inc., where he was working on a variety of acoustical modeling, preprocessing and postprocessing algorithms for audio capture, analysis, control, and presentation. His current interests include 3D audio perception, loudspeaker design, loudspeaker pattern analysis and control, loudness modeling, room simulation, stereo image control and analysis, filter design, speech and audio coding methods, audio and speech testing methodology and execution, and implementation concerns in signal processing. His research interests also include cochlear modeling, masking threshold models, methods of reproducing soundfields either literally or perceptually, microphone and soundfield capture techniques, both actively steered and time-invariant.

He joined DTS, Inc., from his position at Neural Audio when Neural was acquired by DTS. Prior to that, he worked for 6 years at Microsoft Corporation in the Codecs, Core Media Processing and finally the Video Services groups as Audio Architect.

He is retired from AT&T Labs-Research, quartered at Florham Park, NJ, Speech Processing Software and Technology Research Department. Before that, he was employed by AT&T Bell Laboratories, in the Acoustics Research Department under Dr. J. L. Flanagan, and in the Signal Processing Research Department. Until 2002, he was the primary researcher and inventor of AT&T's contributions to the MPEG-2 AAC audio coding algorithm. He also represented AT&T in the ANSI accredited group X3L3.1, and X3L3.1 in the ISO-MPEG-AUDIO (AAC) arena in support of the AAC algorithm. He completed the first perceptual soundfield recordings, which received exceptional reviews from the listener and enthusiast community, before retiring from AT&T.

He started his career working on using analog signal processing to do speech coding (APCM, ADPCM, SBC) for testing of algorithms, sampling rates, and quantizer resolutions. Since then, he has worked in analog signal processing, speech coding, voice privacy, quadrature mirror filter design, and perceptual coding of both audio and images. During this work on perceptual audio coding, he was the primary investigator of the early PXFM audio coder which was reported on at the ASSP Digital Audio Meeting in Mohonk, NY in 1986 and a co-inventor and standards proponent of the ASPEC algorithm, the quality leader in the MPEG-1 audio competition.

During this time, he also did an investigation of coding of still-frame images using a forward-driven perceptual model with Dr. R. J. Safranek, also of AT&T Bell Laboratories. This image coder, called PIC (for Perceptual Image Coder), used very simple techniques to provide state of the art still-image compression.

His major contributions include

The primary inventor and architect for a variety of signal processing algorithms related to room correction, loudness processing, perceptual modeling of audio, audio coding, audio soundfield perception and presentation, and standards and ancillary mathematics and science related to audio issues.

MPEG-2 AAC (Advanced Audio Coding) standard, developed in collaboration with Fraunhofer IIS and other experts in the field of audio compression. MPEG-2 AAC is a reworking of the original AT&T Perceptual Audio Coder (PAC), done with Anibal Ferriera.

Co-invention and standardization of the well-known MP3 algorithm.

Awards and Societies

He is presently a Signal Processing Society Distinguished Lecturer from the IEEE 2011 class, serving as a DL for 2011 and 2012. He was also a distinguished lecturer previously.

In 2006, he was awarded the J. L. Flanagan Signal Processing Field Award from the IEEE Signal Processing Society for his work on creation and standardization of perceptual audio coding.

He was elected a Fellow of the IEEE in 2002.

In February 2001, he received a New Jersey Inventor of the Year award for his contributions to MP3 and audio coding in general.

He became a Senior Member of the IEEE, and received an AT&T Technology Medal and AT&T Standards Award in 1998.

In 1997, JJ was elected a Fellow of the Audio Engineering Society for his work on perceptual coding of audio.

He received his BSEE and MSEE from Carnegie-Mellon University, with side interests in mathematics, audio, radio broadcasting and coherent image signal processing.

About The Graphic Design and Photography

Bob and I wanted to give *iTunes Music* a warm, lively, entertaining look and feel, in tune with its "musical" content. We hope this layout, enriched with informative color graphics and photos, enhances your reading experience.

The photograph on the dedication page reminds me of the legendary Higgs boson, a subatomic particle. A nucleus of thought spreading out into the universe, much like the creative mind of the late Steve Jobs (which still influences our lives for many years to come). It is a photo of Sanford, FL fireworks, taken July 4, 2012. The photo on the credits page is another abstract of the fireworks display.

Our showcase of animals comes from a deep love for these special creatures. We've given them names for the book, but these are not their original or real names. Ajax is a music-loving dog, enjoying that serene feeling we all get when listening to music. The Space Monkeys are here to remind you never to truncate your data or to transcode (or the Space Monkeys will appear!). But they also were early space travelers, little astronauts who impacted our imagination. Many gave their lives unwittingly. Their amazing, charming, innocent expressions belie the dangerous adventures they undertook on our behalf. NASA also gave each monkey his or her name and cared for them as lovingly as possible. For the story of these creatures, visit the NASA website, http://history.nasa.gov/animals.html. We hope you apply the same sense of adventure to your iTunes encodes—think outside the box.

The photo on page 13 is a real church tombstone I found in Orlando, FL, where (apparently) Mark Pause has been buried. Bob says we do miss Mark's presence. The loudness genie's photo on page 64 was taken in Digital Domain's studio B located in Photoshop County, FL (sorry to make you rotate the book, to see him enlarged in all his glory).

Cameras were the Nikon D700 and D800, lenses included Nikon 14-24, 24-70, and Tamron 70-300.

Mary Kent

Index